Cambridge Tracts in Mathematics
and Mathematical Physics

GENERAL EDITORS
H. BASS, J. F. C. KINGMAN, F. SMITHIES, J. A. TODD AND
C. T. C. WALL

No. 55

ASYMPTOTIC EXPANSIONS

ASYMPTOTIC EXPANSIONS

BY

E. T. COPSON

*Regius Professor of Mathematics in
the University of St Andrews*

CAMBRIDGE UNIVERSITY PRESS

CAMBRIDGE

LONDON · NEW YORK · MELBOURNE

PUBLISHED BY THE PRESS SYNDICATE OF THE UNIVERSITY OF CAMBRIDGE
The Pitt Building, Trumpington Street, Cambridge, United Kingdom

CAMBRIDGE UNIVERSITY PRESS
The Edinburgh Building, Cambridge CB2 2RU, UK
40 West 20th Street, New York NY 10011–4211, USA
477 Williamstown Road, Port Melbourne, VIC 3207, Australia
Ruiz de Alarcón 13, 28014 Madrid, Spain
Dock House, The Waterfront, Cape Town 8001, South Africa

http://www.cambridge.org

First published 1965
Reprinted 1967 1971 1976
First paperback edition 2004

A catalogue record for this book is available from the British Library

ISBN 0 521 04721 8 hardback
ISBN 0 521 60482 6 paperback

CONTENTS

Preface   *page* vii

1. Introduction   1

2. Preliminaries   5

3. Integration by parts   13

4. The method of stationary phase   27

5. The method of Laplace   36

6. Watson's lemma   48

7. The method of steepest descents   63

8. The saddle-point method   91

9. Airy's integral   99

10. Uniform asymptotic expansions   107

Bibliography   118

Index   120

PREFACE

In 1943, at the request of the Admiralty Computing Service, I wrote a short monograph on *The Asymptotic Expansion of a Function Defined by a Definite Integral or Contour Integral*. This was one of a series of monographs intended for use in Admiralty Research Establishments, on topics which appeared to be inadequately covered in easily accessible literature. It evidently met a need of the time, since a revised edition was issued in 1946 and had a wide circulation.

The Admiralty monograph has long been unobtainable, and several of my friends have urged me to write this more extensive book on the same general lines. There are few theorems; the aim is the modest one of explaining the methods which are available, and illustrating them by means of a few of the more important special functions.

I must express my thanks to Professor Arthur Erdélyi for the generous advice and encouragement he has given me during the writing of this book.

E. T. C.

Chapter 1

INTRODUCTION

Although the subject of 'Modern Analysis' had its beginnings in the seventeenth century, the mathematicians of the eighteenth century were often more interested in the formal use of infinite processes than in their rigorous proofs. Some of these formal results are very striking.

For example, in 1730, Stirling in his *Methodus Differentialis* gave an infinite series for $\log(m!)$ which, in modern notation, would be written as

$$\log(m!) = z\log z - z + \tfrac{1}{2}\log(2\pi) + \sum_1^\infty \frac{B_{2n}(\tfrac{1}{2})}{(2n-1)(2n)z^{2n-1}},$$

where $z = m + \tfrac{1}{2}$, and $B_n(x)$ is Bernoulli's polynomial defined by

$$\frac{z\,e^{xz}}{e^z - 1} = \sum_0^\infty B_n(x)\frac{z^n}{n!}.$$

Stirling gave the first five coefficients and a recurrence formula for successively determining the coefficients. A similar formula

$$\log(m!) = (m+\tfrac{1}{2})\log m - m + \tfrac{1}{2}\log(2\pi) + \sum_1^\infty \frac{B_{2n}(0)}{(2n-1)(2n)m^{2n-1}}$$

was subsequently given by De Moivre. Both these series are divergent; yet Stirling was able to calculate $\log_{10}(1000!)$, a number between 2567 and 2568, to ten places of decimals by taking only the first few terms of his series. Any partial sum of either of these divergent series is an approximation of $\log(m!)$ with an error of the order of the first term omitted; and, since these terms decrease very rapidly initially, the sum of a few terms may give a very good approximation.

Another interesting result, due to Euler, is that

$$1 + \tfrac{1}{2} + \tfrac{1}{3} + \ldots + \frac{1}{m} - \log m = \gamma + \frac{1}{2m} - \sum_1^\infty \frac{B_{2n}(0)}{(2n)m^{2n}},$$

where γ is Euler's constant. Euler pointed out that the series is divergent for $m = 1$, but he used the series with $m = 10$ to calculate γ correct to 15 places of decimals; yet the series diverges for all values of m. The series is an alternating series, whose terms at first decrease in numerical value; the partial sums lie alternately above and below the desired 'sum', and the accuracy of an approximation by a partial sum can never be better than the least term.

In his *Théorie analytique des probabilités*, published in Paris in 1812, Laplace introduced two new ideas. He showed that the error function

$$\mathrm{Erf}\,(T) = \int_0^T e^{-t^2}\,dt$$

can be represented by a convergent power series

$$\sum_0^\infty (-1)^n \frac{T^{2n+1}}{(2n+1)\,.\,n!},$$

which he called a *série-limite*, because successive partial sums lie alternately above and below the value of the integral, and so provide upper and lower bounds for its value. But when T is large, the series converges so slowly that these bounds are of little use. He also obtained by integration by parts a *série-limite* for the related function

$$\mathrm{Erfc}\,(T) = \int_T^\infty e^{-t^2}\,dt,$$

namely, $\qquad \dfrac{e^{-T^2}}{2T}\left\{1 - \dfrac{1}{2T^2} + \dfrac{1\,.\,3}{(2T^2)^2} - \dfrac{1\,.\,3\,.\,5}{(2T^2)^3} + \ldots\right\}.$

Laplace remarked 'La série a l'inconvénient de finir par être divergente', yet its *série-limite* property made it possible to compute from it values of Erfc (T) for large values of T.

The second idea was that the value of an integral

$$\int \phi(x)\,\{u(x)\}^s\,dx$$

when s is large, depends on the behaviour of $u(x)$ near its stationary points. Laplace used this to obtain the result

$$s! = s^{s+\frac{1}{2}}\,e^{-s}\,\sqrt{(2\pi)}\left[1 + \frac{1}{12s} + \ldots\right],$$

which is frequently, but incorrectly, called Stirling's approximation to the factorial.

Legendre, in his *Traité des fonctions elliptiques* (1825–28), called an infinite series *demi-convergente* if it represented a given function in the sense that the error committed by stopping at any term is of the order of the first term omitted. The term was unfortunate as this property has nothing to do with convergence; for example, the series

$$\sum_{0}^{\infty} (-1)^n r^n \quad (0 < r < 1) \qquad \text{and} \qquad \sum_{1}^{\infty} \frac{1}{n^2}$$

are both convergent, the first being semiconvergent, the second not semiconvergent. Yet the usage has persisted; it occurs, for instance, in Jahnke and Emde's *Funktionentafeln*. A semiconvergent series is nowadays said to be *asymptotic*.

During the nineteenth century, asymptotic expansions were obtained for many of the special functions of analysis, sometimes only formally, sometimes with a rigorous discussion of the order of magnitude of the error. Of particular interest is Stieltjes's [26] doctorate thesis in which he examined the error committed by stopping at the least term of the asymptotic representations of certain important special functions, and showed how the approximation so obtained could be improved.

The modern theory of asymptotic expansions originated in the work of Poincaré [22]. The subject falls roughly into two parts. The first part deals with the summability of asymptotic series, and with the validity of such operations as term by term differentiation or integration; the second is concerned with the actual construction of a series which represents a given function asymptotically.

This tract discusses the asymptotic representation of a function defined by a definite integral or contour integral, usually an analytic function of a complex variable z. If z is complex, we denote its real and imaginary parts by $\mathscr{R}z$ and $\mathscr{I}z$. If z is written in polar form
$$z = r\cos\theta + ir\sin\theta \quad (r > 0),$$
we call θ the phase of z, and write $\theta = \mathrm{ph}\, z$. The phase of z is not unique; it is determined by z only up to an additive multiple of 2π. The principal value of $\mathrm{ph}\, z$ satisfies the inequality

$-\pi < \mathrm{ph}\, z \leqslant \pi$; but since non-integral powers of z will occur, values of $\mathrm{ph}\, z$ other than the principal value turn out to be important.

The terminology and notation used here for the special functions are those of Erdélyi, Magnus, Oberhettinger and Tricomi's three-volume work *Higher Transcendental Functions* (McGraw-Hill, 1953). As this book is very well indexed, no references to it will be given unless it is absolutely necessary. It should be pointed out that the notation of this book is not always the same as that of Whittaker and Watson's *Modern Analysis* (Cambridge, 1920).

The bibliography on pages 118 and 119 is merely a list of works to which reference is made in the text. For example, Poincaré [22] on page 3 refers to Henri Poincaré's *Acta Mathematica* paper which is the twenty-second item in the bibliography.

CHAPTER 2

PRELIMINARIES

1. Asymptotic sequences

Let $f(z)$ and $\phi(z)$ be two functions defined on a set R in the complex plane, and let z_0 be a limit point of R, possibly the point at infinity. For example, R may be a sector

$$0 < |z| < \infty, \quad \alpha < \operatorname{ph} z < \beta,$$

and z_0 might then be the origin or the point at infinity. By a neighbourhood of z_0 (more strictly a spherical neighbourhood), we mean an open disc $|z - z_0| < \delta$ if z_0 is at a finite distance, a region $|z| > \delta$ if z_0 is the point at infinity.

In the usual notation we write $f = O(\phi)$ if there exists a constant A such that $|f| \leqslant A |\phi|$ for all z in R. We also write $f = O(\phi)$ as $z \to z_0$ if there exists a constant A and a neighbourhood U of z_0 such that $|f| \leqslant A |\phi|$ for all points in the intersection of U and R; and $f = o(\phi)$ as $z \to z_0$ if, for any positive number ϵ, there exists a neighbourhood U of z_0 such that $|f| \leqslant \epsilon |\phi|$ for all points z of the intersection of U and R. More simply, if ϕ does not vanish on R, $f = O(\phi)$ means that f/ϕ is bounded, $f = o(\phi)$ that f/ϕ tends to zero as $z \to z_0$.

A sequence of functions $\{\phi_n(z)\}$ is called an *asymptotic sequence* as $z \to z_0$ if there is a neighbourhood of z_0 in which none of the functions vanish (excepting the point z_0) and if for all n

$$\phi_{n+1} = o(\phi_n) \quad \text{as} \quad z \to z_0.$$

For example, if z_0 is finite, $\{(z - z_0)^n\}$ is an asymptotic sequence as $z \to z_0$; and $\{z^{-n}\}$ is as $z \to \infty$.

2. Poincaré's definition of an asymptotic expansion

The formal series

$$\sum_0^\infty a_n \phi_n(z),$$

not necessarily convergent, is said to be an asymptotic expansion of $f(z)$ in Poincaré's sense, with respect to the asymptotic sequence $\{\phi_n(z)\}$ if, for every value of m,

$$f(z) - \sum_0^m a_n\,\phi_n(z) = o(\phi_m(z)),$$

as $z \to z_0$. Since

$$f(z) - \sum_0^{m-1} a_n\,\phi_n(z) = a_m\,\phi_m(z) + o(\phi_m(z)),$$

the partial sum $\qquad\qquad \sum_0^{m-1} a_n\,\phi_n(z)$

is an approximation to $f(z)$ with an error $O(\phi_m)$ as $z \to z_0$; this error is of the same order of magnitude as the first term omitted. If such an asymptotic expansion exists, it is unique, and the coefficients are given successively by

$$a_m = \lim_{z \to z_0} \left\{ f(z) - \sum_0^{m-1} a_n\,\phi_n(z) \right\} \Big/ \phi_m(z).$$

If a function possesses an asymptotic expansion in this sense, we write

$$f(z) \sim \sum_0^\infty a_n\,\phi_n(z).$$

A partial sum of this formal series will often be called an *asymptotic approximation* to $f(z)$. The first term is called the *dominant term*; and we frequently write $f(z) \sim a_0\,\phi_0(z)$, meaning that $f(z)/\phi_0(z)$ tends to a_0 as $z \to z_0$.

The definition has been given for functions of a complex variable z, but it can easily be modified for functions of a real variable x. If the limit point x_0 is finite, R could be an open interval of which x_0 is an internal or end-point; and a neighbourhood of x_0 is an open interval $|x - x_0| < \delta$. But if x_0 is infinite, we have to discriminate between $x \to +\infty$, in which case R could be a semi-infinite interval $x > a$ say, and $x \to -\infty$, in which case R could be, say, $x < b$. There are cases when R is a discrete set; for instance, it might be necessary to find an asymptotic expansion for the nth partial sum of an infinite series when n is large, but such problems lie, in the main, outside the scope of this tract.

The form of an asymptotic expansion evidently depends on the choice of the asymptotic sequence. For example, as $z \to \infty$,

$$\frac{1}{z-1} \sim \sum_1^\infty \frac{1}{z^n},$$

and

$$\frac{1}{z-1} \sim \sum_1^\infty \frac{z+1}{z^{2n}}.$$

In these examples it happens that the asymptotic expansions are convergent series.

Again, two functions may have the same asymptotic expansion. For example, if $-\frac{1}{2}\pi + \delta \leqslant \mathrm{ph}\, z \leqslant \frac{1}{2}\pi - \delta$, where $0 < \delta < \frac{1}{2}\pi$, the two functions

$$\frac{1}{z+1}, \quad \frac{1}{z+1} + e^{-z},$$

both have the asymptotic expansion

$$\sum_1^\infty \frac{(-1)^{n-1}}{z^n}$$

as $z \to \infty$, since $z^n e^{-z}$ tends to zero as $z \to \infty$ in the given sector.

3. Asymptotic power series

If the limit point z_0 is at a finite distance, it is transformed into the point at infinity by $z' = 1/(z - z_0)$. We shall suppose that this has been done and consider only asymptotic expansions as $z \to \infty$ in a sector $\alpha < \mathrm{ph}\, z < \beta$; or, in the case of a function of a real variable x, as $x \to +\infty$ or as $x \to -\infty$.

The simplest type of asymptotic sequence as $z \to \infty$ is $\{\phi(z)/z^n\}$. If a function $f(z)$ possesses an asymptotic expansion with respect to this sequence, say

$$f(z) \sim \phi(z) \sum_0^\infty \frac{a_n}{z^n},$$

this implies that

$$\frac{f(z)}{\phi(z)} \sim \sum_0^\infty \frac{a_n}{z^n},$$

the latter series being an asymptotic expansion with respect to the sequence $\{1/z^n\}$. An asymptotic expansion with respect to the sequence $\{1/z^n\}$ is called an *asymptotic power series*.

4. Calculations with asymptotic power series

Asymptotic power series and convergent power series possess very similar formal properties. The main results are stated below, first for the case of a real variable. It is assumed that $f(x)$ and $g(x)$ possess asymptotic expansions

$$f(x) \sim \sum_0^\infty \frac{a_n}{x^n}, \quad g(x) \sim \sum_0^\infty \frac{b_n}{x^n}$$

as $x \to +\infty$.

(i) *If A is a constant,*

$$Af(x) \sim \sum_0^\infty \frac{Aa_n}{x^n},$$

(ii) $$f(x)+g(x) \sim \sum_0^\infty \frac{a_n+b_n}{x^n}.$$

These results follow at once from the definition.

(iii) $$f(x)\,g(x) \sim \sum_0^\infty \frac{c_n}{x^n},$$

where $\quad c_n = a_0\,b_n + a_1\,b_{n-1} + \ldots + a_{n-1}\,b_1 + a_n\,b_0.$

For any positive integer N,

$$f(x) = a_0 + \frac{a_1}{x} + \ldots + \frac{a_N}{x^N} + O\!\left(\frac{1}{x^{N+1}}\right),$$

$$g(x) = b_0 + \frac{b_1}{x} + \ldots + \frac{b_N}{x^N} + O\!\left(\frac{1}{x^{N+1}}\right),$$

and hence $\quad f(x)\,g(x) = c_0 + \dfrac{c_1}{x} + \ldots + \dfrac{c_N}{x^N} + O\!\left(\dfrac{1}{x^{N+1}}\right),$

which was to be proved.

It follows that any positive integral power of $f(x)$ possesses an asymptotic power series expansion, and so also does any polynomial in $f(x)$.

(iv) *If $a_0 \neq 0$, then*

$$\frac{1}{f(x)} \sim \frac{1}{a_0} + \sum_1^\infty \frac{d_n}{x^n},$$

as $x \to +\infty$.

In the first place $1/f(x)$ tends to a finite limit $1/a_0$ as $x \to \infty$.
Next
$$\left\{\frac{1}{f(x)} - \frac{1}{a_0}\right\} \bigg/ \frac{1}{x} = x\left\{\frac{1}{a_0 + (a_1/x) + O(1/x^2)} - \frac{1}{a_0}\right\}$$

$$= \frac{-a_1 + O(1/x)}{a_0\{a_0 + (a_1/x) + O(1/x^2)\}} \to -\frac{a_1}{a_0^2}.$$

Similarly,
$$\left\{\frac{1}{f(x)} - \frac{1}{a_0} + \frac{a_1}{a_0^2 x}\right\} \bigg/ \frac{1}{x^2} \to \frac{a_1^2 - a_0 a_2}{a_0^3},$$

and so on. The successive coefficients d_n are found in a similar way.

More generally, any rational function of $f(x)$ has an asymptotic power series expansion provided that the denominator does not tend to zero as $x \to +\infty$.

(v) *If $f(x)$ is continuous when $x > a > 0$, then, if $x > a$,*

$$F(x) = \int_x^\infty \left\{f(t) - a_0 - \frac{a_1}{t}\right\} dt$$

has the asymptotic power series expansion

$$F(x) \sim \frac{a_2}{x} + \frac{a_3}{2x^2} + \ldots + \frac{a_{n+1}}{nx^n} + \ldots$$

as $x \to +\infty$.

Since $f(t) - a_0 - a_1/t$ is continuous when $t > a$ and is $O(1/t^2)$ as $t \to +\infty$, the integral $F(x)$ exists for $x > a$.

Since
$$F(x) = \int_x^\infty \left\{\sum_2^m \frac{a_n}{t^n} + O\left(\frac{1}{t^{m+1}}\right)\right\} dt$$

for every integer $m \geqslant 2$, we have

$$F(x) = \sum_2^m \frac{a_n}{(n-1)x^{n-1}} + O\left(\frac{1}{x^m}\right)$$

$$= \sum_1^{m-1} \frac{a_{n+1}}{nx^n} + O\left(\frac{1}{x^m}\right)$$

as $x \to +\infty$; and the result follows.

(vi) *If $f(x)$ has a continuous derivative $f'(x)$, and if $f'(x)$ possesses an asymptotic power series expansion as $x \to +\infty$, the latter expansion is*

$$f'(x) \sim -\sum_2^\infty \frac{(n-1)a_{n-1}}{x^n}.$$

B

For suppose that
$$f'(x) \sim \sum_0^\infty \frac{b_n}{x^n},$$

as $x \to \infty$. Now, since $f'(x)$ is continuous,

$$f(y) - f(x) = \int_x^y f'(t)\,dt$$
$$= b_0(y-x) + b_1 \log \frac{y}{x} + \int_x^y \left\{ f'(t) - b_0 - \frac{b_1}{t} \right\} dt.$$

But $f(y) \to a_0$ as $y \to +\infty$, and

$$\int_x^\infty \left\{ f'(t) - b_0 - \frac{b_1}{t} \right\} dt$$

is convergent since the integrand is $O(1/t^2)$. It follows that $b_0 = b_1 = 0$ and that

$$a_0 - f(x) = \int_x^\infty f'(t)\,dt.$$

By (v),
$$a_0 - f(x) \sim \sum_1^\infty \frac{b_{n+1}}{nx^n} \quad \text{as} \quad x \to +\infty.$$

But we know that
$$a_0 - f(x) \sim -\sum_1^\infty \frac{a_n}{x^n}.$$

Since an asymptotic power-series expansion is unique,

$$b_{n+1} = -na_n,$$

that is
$$f'(x) \sim -\sum_2^\infty \frac{(n-1)\,a_{n-1}}{x^n},$$

as $x \to +\infty$. In other words, the asymptotic expansion is obtained by formal term by term differentiation.

These results have been stated for functions of a real variable x as $x \to +\infty$. They could have been stated, almost word for word, for functions of a complex variable z as $z \to \infty$ either in a sector or in a whole neighbourhood of the point at infinity.

(vi) can be modified in the case of analytic functions of a complex variable z which are, by definition, differentiable. The result in this case is:

(vii) *If $f(z)$ is an analytic function, regular in the region R defined by $|z| > a$, $\alpha < |\mathrm{ph}\,z| < \beta$, and if*

$$f(z) \sim a_0 + \frac{a_1}{z} + \frac{a_2}{z^2} + \dots$$

uniformly in ph z *as* $|z| \to \infty$ *in any closed sector contained in* R, *then*

$$f'(z) \sim -\frac{a_1}{z^2} - \frac{2a_2}{z^3} - \frac{3a_3}{z^4} - \cdots$$

uniformly in ph z *as* $|z| \to \infty$ *in any closed sector contained in* R.

When we say that the asymptotic power series expansion of $f(z)$ holds uniformly in ph z as $|z| \to \infty$ in a closed sector

$$\alpha_1 \leqslant \text{ph} z \leqslant \beta_1$$

contained in R, we mean that, for every integer m,

$$f(z) = \sum_{n=0}^{m-1} \frac{a_n}{z^n} + \frac{\phi_m(z)}{z^m},$$

where $\phi_m(z)$ is bounded in $|z| \geqslant a_1$, $\alpha_1 \leqslant \text{ph} z \leqslant \beta_1$, that is, for each integer m, there exists a constant A_m such that

$$|\phi_m(z)| < A_m$$

there.

Since $f(z)$ is regular in R, it is, by definition, differentiable, and hence $\phi_m(z)$ is regular in R, and

$$f'(z) = -\sum_{n=1}^{m-1} \frac{na_n}{z^{n+1}} + \frac{\phi_m'(z)}{z^m} - \frac{m\phi_m(z)}{z^{m+1}}$$

$$= -\sum_{n=2}^{m-1} \frac{(n-1)a_{n-1}}{z^n} + \frac{\psi_m(z)}{z^m},$$

where $\qquad \psi_m(z) = \phi_m'(z) - (m-1)a_{m-1} - \dfrac{m\phi_m(z)}{z}.$

We have to show that $\psi_m(z)$ is bounded in any closed sector $\alpha_2 \leqslant \text{ph} z \leqslant \beta_2$, contained in R. Evidently it suffices to show that $\phi_m'(z)$ is bounded.

Given α_2, β_2, we choose α_1 and β_1 so that

$$\alpha < \alpha_1 < \alpha_2 < \beta_2 < \beta_1 < \beta.$$

Then $|\phi_m(z)| < A_m$ in $\alpha_1 \leqslant \text{ph} z \leqslant \beta_1$. We can choose a positive number δ so that, if ζ lies in $\alpha_2 \leqslant \text{ph} \zeta \leqslant \beta_2$, the circle c whose equation is $|z - \zeta| = \delta|\zeta|$ lies in $\alpha_1 \leqslant \text{ph} z \leqslant \beta_1$. Hence

$$\phi_m'(\zeta) = \frac{1}{2\pi i} \int_c \frac{\phi_m(z)}{(z-\zeta)^2} dz = \frac{1}{2\pi} \int_0^{2\pi} \frac{\phi_m(\zeta + \delta\zeta e^{\theta i})}{\delta\zeta e^{\theta i}} d\theta,$$

and so $\qquad\qquad |\phi_m'(\zeta)| \leqslant \dfrac{A_m}{\delta|\zeta|} \leqslant \dfrac{A_m}{\delta a}.$

This proves the result.

An asymptotic power series expansion of an analytic function usually holds in a sectorial region. Such a function may possess different asymptotic expansions in different sectors, an effect known as the Stokes Phenomenon.

(viii) *If $f(z)$ is a one-valued function regular in $|z| \geqslant a$ and if*

$$f(z) \sim \sum_{0}^{\infty} \frac{a_n}{z^n}$$

as $z \to \infty$ for all values of ph z, *the asymptotic power series is convergent with sum $f(z)$ for all sufficiently large values of $|z|$.*

Let R_1 be any number greater than a. Then $f(z)$ has a Laurent expansion

$$f(z) = \sum_{-\infty}^{\infty} c_n z^n$$

convergent in $|z| \geqslant R_1$, where

$$c_n = \frac{1}{2\pi i} \int_{\Gamma} \frac{f(z)}{z^{n+1}} dz,$$

Γ being any circle $|z| = R$ where $R > R_1$. Since $f(z)$ tends to a_0 as $z \to \infty$, it is bounded; thus there exists a constant M such that $|f(z)| \leqslant M$ when $|z| \geqslant a$. When $n > 0$,

$$|c_n| \leqslant \frac{M}{R^n}.$$

But R can be as large as we please. Hence $c_n = 0$ when $n > 0$, and therefore

$$f(z) = \sum_{0}^{\infty} c_{-n} z^{-n},$$

the series being convergent when $|z| \geqslant R_1$. But

$$f(z) \sim \sum_{0}^{\infty} a_n z^{-n}$$

as $z \to \infty$, and so

$$a_0 = \lim_{z \to \infty} f(z) = c_0,$$

$$a_1 = \lim_{z \to \infty} \{f(z) - a_0\} \Big/ \frac{1}{z} = c_{-1},$$

$$a_2 = \lim_{z \to \infty} \left\{ f(z) - a_0 - \frac{a_1}{z} \right\} \Big/ \frac{1}{z^2} = c_{-2};$$

in general $a_n = c_{-n}$. Hence the asymptotic power series expansion of $f(z)$ is convergent.

CHAPTER 3

INTEGRATION BY PARTS

5. The Incomplete Gamma Function

One of the simplest ways of finding the asymptotic expansion of a function defined by a definite integral is the method of integration by parts. The successive terms of the asymptotic series are produced by repeated integration by parts, and the asymptotic character of the series is then proved by examining the remainder, which is in the form of a definite integral. The field of application of the method is rather restricted, and it is difficult to formulate precise theorems of any degree of generality. Instead of attempting this, we try to make the idea clear by discussing particular examples.

As a first example, we take the Incomplete Gamma Function defined by

$$\gamma(a,x) = \int_0^x e^{-t}t^{a-1}\,dt,$$

where x and a are positive. A series suitable for calculation when x is small can be deduced at once by expanding the exponential function and integrating term by term. This series

$$\gamma(a,x) = \sum_0^\infty \frac{(-1)^n}{a+n}\frac{x^{n+a}}{n!}$$

converges for all positive values of x, but is of little use for numerical work. For example, if $x = 10$, $a = \frac{1}{2}$, the largest term, corresponding to $n = 8$, is about 923, yet $\gamma(\frac{1}{2}, 10)$ is equal to $\sqrt{\pi}$ with an error of the order of 10^{-5}.

When x is large and positive, it is better to consider the function

$$\Gamma(a,x) = \Gamma(a) - \gamma(a,x) = \int_x^\infty e^{-t}t^{a-1}\,dt,$$

where the integral is convergent for all values of the parameter a. If we integrate by parts once, we get

$$\Gamma(a,x) = e^{-x}x^{a-1} + (a-1)\,\Gamma(a-1,x).$$

If we repeat the process, we find that, if a is a positive integer, $\Gamma(a, x)$ is the product of e^{-x} and a polynomial in x of degree $a - 1$. But, in general, we obtain after n integrations by parts

$$\Gamma(a, x) = \sum_{r=1}^{n} \frac{\Gamma(a)}{\Gamma(a-r+1)} e^{-x} x^{a-r} + \frac{\Gamma(a)}{\Gamma(a-n)} \Gamma(a-n, x).$$

Now
$$\left| \frac{\Gamma(a)}{\Gamma(a-n)} \int_{x}^{\infty} e^{-t} t^{a-n-1} dt \right|$$

$$< \left| \frac{\Gamma(a)}{\Gamma(a-n)} \right| x^{a-n-1} \int_{x}^{\infty} e^{-t} dt$$

$$= \left| \frac{\Gamma(a)}{\Gamma(a-n)} \right| e^{-x} x^{a-n-1}$$

if $n > a - 1$. Hence as $x \to +\infty$,

$$\Gamma(a, x) \sim \sum_{r=1}^{\infty} \frac{\Gamma(a)}{\Gamma(a-r+1)} e^{-x} x^{a-r},$$

since the error in stopping at the nth term is less in absolute value than the first term omitted.

A particular case of this result is the asymptotic expansion of the error function

$$\text{Erfc } T = \int_{T}^{\infty} e^{-u^2} du = \tfrac{1}{2} \Gamma(\tfrac{1}{2}, T^2),$$

as $T \to +\infty$, namely

$$\text{Erfc } T \sim \tfrac{1}{2} \sqrt{\pi}\, e^{-T^2} \sum_{r=1}^{\infty} \frac{1}{\Gamma(\tfrac{3}{2} - r)\, T^{2r-1}},$$

which readily transforms into Laplace's result

$$\text{Erfc } T \sim \frac{1}{2\sqrt{\pi}} e^{-T^2} \sum_{r=1}^{\infty} \Gamma(r - \tfrac{1}{2}) \frac{(-1)^{r-1}}{T^{2r-1}}.$$

6. The Fresnel and allied integrals

The Fresnel integrals of physical optics

$$\int_{u}^{\infty} \cos(\theta^2)\, d\theta, \quad \int_{u}^{\infty} \sin(\theta^2)\, d\theta$$

may be written in the forms

$$\int_{u^2}^{\infty} \frac{\cos t}{\sqrt{t}}\, dt, \quad \int_{u^2}^{\infty} \frac{\sin t}{\sqrt{t}}\, dt.$$

They are particular cases of the real and imaginary parts of the integral

$$F(x,a) = \int_{x}^{\infty} \frac{e^{it}}{t^a}\, dt,$$

which converges for all positive values of x if a is positive.

If we integrate by parts, we obtain

$$F(x,a) = \frac{i\,e^{ix}}{x^a} - ia\,F(x, a+1).$$

Repeating this argument, we find that

$$F(x,a) = \frac{i\,e^{ix}}{x^a} \sum_{r=0}^{n} \frac{\Gamma(a+r)}{\Gamma(a)\,(ix)^r} + \frac{1}{i^{n+1}} \frac{\Gamma(a+n+1)}{\Gamma(a)} F(x, a+n+1).$$

From this it follows that

$$F(x,a) \sim \frac{i\,e^{ix}}{x^a} \sum_{r=0}^{\infty} \frac{\Gamma(a+r)}{\Gamma(a)\,(ix)^r}$$

as $x \to +\infty$; for the absolute value of the remainder after $n+1$ terms is

$$\frac{\Gamma(a+n+1)}{\Gamma(a)} \left| \int_{x}^{\infty} \frac{e^{it}}{t^{a+n+1}}\, dt \right| \leqslant \frac{\Gamma(a+n+1)}{\Gamma(a)} \int_{x}^{\infty} \frac{dt}{t^{a+n+1}}$$

$$= \frac{\Gamma(a+n)}{\Gamma(a)\, x^{a+n}},$$

which is the absolute value of the $(n+1)$th term. Hence the remainder after n terms does not exceed in absolute value the absolute value of the $(n+1)$th term, which proves the result. This asymptotic expansion evidently also holds when a is a complex constant with positive real part.

7. A problem of Stieltjes

In his thesis Stieltjes [26] applied the method of integration by parts to the function

$$F(z) = \int_{0}^{\infty} e^{-t} \frac{dt}{t+z}. \tag{7.1}$$

Since the integral converges uniformly in

$$|z| \geqslant \epsilon > 0, \quad |\mathrm{ph}\, z| \leqslant \pi - \delta < \pi,$$

$F(z)$ is an analytic function, regular in the complex plane cut along the negative real axis. It is the confluent hypergeometric function $\Psi(1, 1; z)$, whose only singularity is a branch point at the origin near which it behaves like the principal value of $\log 1/z$.

If we integrate by parts n times, we obtain

$$F(z) = \sum_{r=1}^{n} \frac{(-1)^{r-1}(k-1)!}{z^r} + (-1)^n n! \int_0^\infty e^{-t} \frac{dt}{(t+z)^{n+1}}.$$

The divergent series
$$\sum_{r=1}^{\infty} \frac{(-1)^{r-1}(r-1)!}{z^r}$$

is an asymptotic power series expansion of $F(z)$, valid as $|z| \to \infty$ in $|\mathrm{ph}\, z| \leqslant \pi - \delta$.

To prove this, we observe that when $t \geqslant 0$ and $|\mathrm{ph}\, z| \leqslant \pi - \delta$, then $|z+t| \geqslant |z| \sin \delta$. The absolute value of the remainder $R_n(z)$ after n terms is

$$|R_n(z)| = \left| (-1)^n n! \int_0^\infty e^{-t} \frac{dt}{(t+z)^{n+1}} \right| \leqslant \frac{n!}{|z|^{n+1} \sin^{n+1} \delta} \int_0^\infty e^{-t} dt$$

$$= \frac{n!}{|z|^{n+1} \sin^{n+1} \delta},$$

which is of the same order of magnitude, as $|z| \to \infty$, as the $(n+1)$th term. Hence

$$F(z) \sim \sum_{r=1}^{\infty} \frac{(-1)^{r-1}(r-1)!}{z^r}. \qquad (7.2)$$

When $z = x > 0$,

$$0 < (-1)^n R_n(x) = n! \int_0^\infty e^{-t} \frac{dt}{(t+x)^{n+1}} < \frac{n!}{x^{n+1}},$$

so that the remainder after n terms has the same sign as the $(n+1)$th term and is less than it in absolute value. This asymptotic expansion is an alternating series, a *série-limite* in Laplace's terminology, in that successive partial sums lie alternately above and below the value of $F(x)$. The best approximation by a partial sum is obtained by stopping just short of the smallest term. Stieltjes showed that a still better approximation can then be obtained by adding half the smallest term.

If $x = N + \eta$, where N is a positive integer and $0 \leqslant \eta < 1$, the smallest term is the $(N+1)$th term. The remainder after N terms is

$$R_N(N+\eta) = (-1)^N N! \int_0^\infty e^{-t} \frac{dt}{(t+N+\eta)^{N+1}}$$

$$= \frac{(-1)^N N!}{(N+\eta)^{N+1}} \int_0^\infty e^{-t} \left(1 + \frac{t}{N+\eta}\right)^{-N-1} dt.$$

The integrand is dominated by e^{-t}. By the dominated convergence theorem,

$$R_N(N+\eta) \bigg/ \frac{(-1)^N N!}{(N+\eta)^{N+1}} \to \int_0^\infty e^{-2t} dt = \tfrac{1}{2}$$

as $N \to \infty$. Hence if $x = N + \eta$, where N is a large integer and $0 \leqslant \eta < 1$, the remainder after N terms is approximately one half of the $(N+1)$th term, that is, one half of the smallest term.

8. The analytical continuation of Stieltjes's function

The function $F(z)$, defined in $|\mathrm{ph}\, z| < \pi$ by equation (7.1) is the principal branch of the confluent hypergeometric function $\Psi(1, 1; z)$ which has a logarithmic branch-point at the origin. Other branches can be obtained by analytical continuation in the following way.

When $\mathscr{R}z > 0$, rotation of the path of integration through a right angle gives the integral

$$F_1(z) = \int_0^\infty e^{-it} \frac{i\, dt}{it+z} \qquad (8.1)$$

as an alternative representation of $F(z)$ in the right-hand half-plane. But $F_1(z)$ is an analytic function, regular in $|z| > 0$, $-\tfrac{1}{2}\pi < \mathrm{ph}\, z < \tfrac{3}{2}\pi$, since the integral in (8.1) converges uniformly on any compact set in this sector. Thus, while $F(z)$ and $F_1(z)$ are the same function in $-\tfrac{1}{2}\pi < \mathrm{ph}\, z < \pi$, $F_1(z)$ provides the analytical continuation of $F(z)$ across the cut $\mathrm{ph}\, z = \pi$.

In the third quadrant, $F(z)$ and $F_1(z)$ are different functions. For

$$F(z) - F_1(z) = \int_0^\infty e^{-s} \frac{ds}{z+s} - \int_0^\infty e^{-it} \frac{i\, dt}{z+it}$$

$$= \lim_{R \to \infty} \int_C e^{-\sigma} \frac{d\sigma}{z+\sigma}, \qquad (8.2)$$

where C is the boundary of the quadrant $|\sigma| \leqslant R$, $0 \leqslant \mathrm{ph}\,\sigma \leqslant \frac{1}{2}\pi$ in the complex σ plane. When z is in the third quadrant, the pole $\sigma = -z$ lies inside C when $R > |z|$, and so

$$F(z) - F_1(z) = 2\pi i\, e^z.$$

Thus
$$F_1(z) = F(z) - 2\pi i\, e^z \qquad (8.3)$$

connects the two branches in the third quadrant.

Although $F(z)$ is discontinuous across the negative real axis, $F_1(z)$ is continuous, and hence, if $a > 0$,

$$F(-a+0i) - F(-a-0i) = -2\pi i\, e^{-a}.$$

Stieltjes observed that both these limiting values can be expressed in terms of a Cauchy principal value integral. For if $a > 0$ and if Γ is the contour of (8.2) indented upwards at $\sigma = a$, we have

$$\lim_{R \to \infty} \int_\Gamma e^{-\sigma}\, \frac{d\sigma}{\sigma - a} = 0.$$

This gives
$$P\int_0^\infty e^{-s}\, \frac{ds}{s-a} - \pi i\, e^{-a} - \int_0^\infty e^{-it}\, \frac{i\, dt}{it - a} = 0,$$

or
$$F_1(-a) = P\int_0^\infty e^{-s}\, \frac{ds}{s-a} - \pi i\, e^{-a}. \qquad (8.4)$$

Since
$$F(-a+0i) = F_1(-a),$$
it follows that

$$F(-a \pm 0i) = P\int_0^\infty e^{-s}\, \frac{ds}{s-a} \mp \pi i\, e^{-a}. \qquad (8.5)$$

The asymptotic expansion of $F_1(z)$ can be obtained at once from (8.1) by integration by parts. The result is that

$$F_1(z) \sim \sum_1^\infty \frac{(-1)^{k-1}(k-1)!}{z^k}$$

as $|z| \to \infty$ in $-\frac{1}{2}\pi + \delta \leqslant \mathrm{ph}\,z \leqslant \frac{3}{2}\pi - \delta < \frac{3}{2}\pi$. Hence, by (8.4),

$$P\int_0^\infty e^{-s}\, \frac{ds}{s-a} \sim -\sum_1^\infty \frac{(k-1)!}{a^k} \qquad (8.6)$$

as $a \to +\infty$, the term involving e^{-a} being omitted since it is very small compared with any term of the series (8.6).

The asymptotic expansion (8.6) is a series of positive terms, not an alternating series, and this fact, as Stieltjes pointed out, makes it more difficult to estimate the best approximation which can be made by taking partial sums.

9. Stieltjes's best approximation

Let us write

$$G(a) = P\int_0^\infty e^{-s}\frac{ds}{a-s}.$$ (9.1)

We have seen that

$$G(a) \sim \sum_1^\infty \frac{(k-1)!}{a^k}$$ (9.2)

as $a \to +\infty$. It is possible to obtain this result in a different way, which leads to a more suitable formula for the remainder after n terms in the asymptotic expansion.

If we write

$$G(a) = P\int_{-\infty}^a e^{t-a}\frac{dt}{t},$$

we have

$$e^a G(a) - e^b G(b) = \int_b^a e^t \frac{dt}{t},$$

when $a > b > 0$, which has the advantage of not involving a Cauchy principal value. Integration by parts n times gives

$$e^a G(a) - e^b G(b) = \sum_{k=1}^n (k-1)!\left\{\frac{e^a}{a^k}-\frac{e^b}{b^k}\right\}+n!\int_b^a \frac{e^v}{v^{n+1}}dv.$$

Therefore, if

$$G(a) = \sum_1^n \frac{(k-1)!}{a^k}+R_n(a),$$

then

$$e^a R_n(a) - e^b R_n(b) = n!\int_b^a \frac{e^v}{v^{n+1}}dv.$$ (9.3)

Now

$$R_n(a) = e^{1-a}R_n(1)+n!\int_1^a \frac{e^{v-a}}{v^{n+1}}dv;$$

we shall show that the right-hand side is $O(1/a^{n+1})$ as $a \to \infty$. This is true of the first term; and, if $a > 2$,

$$0 < \int_1^a \frac{e^{v-a}}{v^{n+1}}\, dv = \int_1^{\frac{1}{2}a} \frac{e^{v-a}}{v^{n+1}}\, dv + \int_{\frac{1}{2}a}^a \frac{e^{v-a}}{v^{n+1}}\, dv$$

$$\leqslant \int_1^{\frac{1}{2}a} e^{v-a}\, dv + \frac{1}{(\frac{1}{2}a)^{n+1}} \int_{\frac{1}{2}a}^a e^{v-a}\, dv$$

$$= \{e^{-\frac{1}{2}a} - e^{1-a}\} + \frac{1}{(\frac{1}{2}a)^{n+1}} \{1 - e^{-\frac{1}{2}a}\} = O\!\left(\frac{1}{a^{n+1}}\right).$$

This proves that (9.2) is the asymptotic expansion of $G(a)$; but the relation (9.3) enables us to get more precise information.

From the particular case

$$e^a R_1(a) - e^b R_1(b) = \int_b^a \frac{e^v}{v^2}\, dv,$$

it follows that $e^x R_1(x)$ increases steadily from $-\infty$ to $+\infty$ as x increases from 0 to $+\infty$. Hence $e^a R_1(a)$ is positive for all sufficiently large values of a. Again, since

$$R_n(a) = \frac{n!}{a^{n+1}} + R_{n+1}(a),$$

$R_n(a)$ decreases steadily from the positive value $R_1(a)$ as n increases. It cannot tend to a finite limit, since $\sum_1^\infty (k-1)!/a^k$ is divergent; hence $R_n(a)$ tends steadily to $-\infty$. There is therefore an integer N with the property $R_N(a) \geqslant 0$, $R_{N+1}(a) \leqslant 0$, that is, the remainder $R_n(a)$ changes sign as n passes through the value N. The best approximation by a partial sum is obtained by taking the partial sum up to the point where $R_n(a)$ changes sign.

Stieltjes proved that this change of sign occurs near the smallest term. If $a = N + \eta$, where N is a large integer and $0 \leqslant \eta < 1$, he showed that

$$R_{N-1}(N+\eta) = \frac{N!}{(N+\eta)^{N+1}} \left\{\eta + \tfrac{2}{3} + O\!\left(\frac{1}{N}\right)\right\},$$

$$R_N(N+\eta) = \frac{N!}{(N+\eta)^{N+1}} \left\{\eta - \tfrac{1}{3} + O\!\left(\frac{1}{N}\right)\right\},$$

$$R_{N+1}(N+\eta) = \frac{N!}{(N+\eta)^{N+1}} \left\{\eta - \tfrac{4}{3} + O\!\left(\frac{1}{N}\right)\right\},$$

so that $R_{N-1}(N+\eta)$ and $R_{N+1}(N+\eta)$ have opposite signs; the sign of $R_N(N+\eta)$ depends on the magnitude of η. We return to this in chapter 6.

For the developments of these ideas, which are of great importance in the practical use of asymptotic expansions in computing, we refer to Dingle's series of papers on 'Asymptotic expansions and converging factors' (Dingle [11]).

10. Fourier integrals

In the previous examples in this chapter, it was possible to obtain a complete asymptotic expansion by integrating by parts an indefinite number of times. There are cases when it is possible to integrate by parts only a finite number of times, and the process then leads to a finite expansion together with an error term. The simplest such case is the Fourier integral over a finite range.

Let $\phi(x)$ be N times continuously differentiable in $\alpha \leqslant x \leqslant \beta$. Then

$$\int_\alpha^\beta e^{i\nu x}\phi(x)\,dx = \sum_{n=0}^{N-1} \frac{i^{n+1}}{\nu^{n+1}}\{e^{i\nu\alpha}\phi^{(n)}(\alpha) - e^{i\nu\beta}\phi^{(n)}(\beta)\} + o\left(\frac{1}{\nu^N}\right) \quad (10.1)$$

as $\nu \to +\infty$, where $\phi^{(n)}(x)$ denotes $d^n\phi/dx^n$.

For if we integrate by parts N times, we obtain

$$\int_\alpha^\beta e^{i\nu x}\phi(x)\,dx = \sum_{n=0}^{N-1} \frac{i^{n+1}}{\nu^{n+1}}\{e^{i\nu\alpha}\phi^{(n)}(\alpha) - e^{i\nu\beta}\phi^{(n)}(\beta)\} + R_N,$$

where $$R_N = \frac{i^N}{\nu^N}\int_\alpha^\beta e^{i\nu x}\phi^{(N)}(x)\,dx.$$

Since $\phi^{(N)}(x)$ is, by hypothesis, continuous, the last integral tends to zero as $\nu \to \infty$ by Riemann's lemma, and so $R_N = o(1/\nu^N)$.

The result remains true if $\beta = \infty$, provided that $\phi^{(n)}(x)$ tends to zero as $x \to +\infty$ for $n = 0, 1, 2, ..., N-1$ and provided that $|\phi^{(N)}(x)|$ is integrable over $x \geqslant \alpha$. And similarly if the lower limit is infinite.

If $\phi(t)$ or its derivatives from some point on have singularities, the argument fails. If there are a finite number of singularities, we need only consider the case when the singularities are at the ends of the range of integration. The simplest case is an integral of the form

$$\int_\alpha^\beta e^{i\nu x}(x-\alpha)^{\lambda-1}(\beta-x)^{\mu-1}\phi(x)\,dx, \qquad (10.2)$$

where $0 < \lambda < 1$, $0 < \mu < 1$ and $\phi(x)$ is N times continuously differentiable.

11. The singular case

In order to apply the process of integration by parts to (10.2) and introduce successive derivatives of $\phi(x)$, it is necessary to have formulae for the successive integrals of

$$e^{i\nu x}(x-\alpha)^{\lambda-1}(\beta-x)^{\mu-1}.$$

The arguments used here I owe to Erdélyi [12].

Let us denote by $If(x)$ the integral of $f(t)$ from a to x. Then

$$I^n f(x) = \frac{1}{(n-1)!}\int_a^x (x-t)^{n-1}f(t)\,dt. \qquad (11.1)$$

The result is true for $n = 1$. If it is true for $n = m$, then

$$I^{m+1}f(x) = \frac{1}{(m-1)!}\int_a^x du\int_a^u (u-t)^{m-1}f(t)\,dt$$

$$= \frac{1}{(m-1)!}\int_a^x dt\int_t^x (u-t)^{m-1}f(t)\,du$$

$$= \frac{1}{m!}\int_a^x (x-t)^m f(t)\,dt,$$

so that the result is true for $n = m+1$. Equation (11.1) then follows by induction.

The lower limit a is an arbitrary constant. If

$$f(x) = e^{i\nu x}(x-\alpha)^{\lambda-1}, \qquad (11.2)$$

where $\nu > 0$, $0 < \lambda < 1$, it is convenient to integrate in the plane of a complex variable σ and take $a = \infty i$. Then

$$I^n f(x) = \frac{1}{(n-1)!}\int_{\infty i}^x (x-\sigma)^{n-1}(\sigma-\alpha)^{\lambda-1}e^{i\nu\sigma}\,d\sigma,$$

where $x \geqslant \alpha$ and $(\sigma - \alpha)^{\lambda - 1}$ has its principal value. If we put $\sigma = x + it$, it follows that

$$I^n f(x) = \frac{e^{i\nu x - \frac{1}{2}n\pi i}}{(n-1)!} \int_0^\infty e^{-\nu t} t^{n-1} (x - \alpha + it)^{\lambda - 1} \, dt,$$

and, in particular,

$$I^n f(\alpha) = \frac{e^{i\nu\alpha + \frac{1}{2}\lambda\pi i - \frac{1}{2}(n+1)\pi i}}{(n-1)!} \int_0^\infty e^{-\nu t} t^{n+\lambda - 2} \, dt$$

$$= e^{i\nu\alpha + \frac{1}{2}\lambda\pi i - \frac{1}{2}(n+1)\pi i} \frac{\Gamma(n+\lambda-1)}{(n-1)! \, \nu^{n+\lambda-1}}. \qquad (11.3)$$

Moreover,

$$|I^n f(x)| \leqslant \frac{1}{(n-1)!} \int_0^\infty e^{-\nu t} t^{n-1} \{(x-\alpha)^2 + t^2\}^{\frac{1}{2}(\lambda-1)} \, dt$$

$$\leqslant \frac{(x-\alpha)^{\lambda-1}}{(n-1)!} \int_0^\infty e^{-\nu t} t^{n-1} \, dt = \frac{(x-\alpha)^{\lambda-1}}{\nu^n}. \qquad (11.4)$$

We are now in a position to prove the following result:

Let $\phi(x)$ be N times continuously differentiable in $\alpha \leqslant x \leqslant \beta$. Let $\phi(x)$ and its first $N-1$ derivatives vanish when $x = \beta$. Then, if $0 < \lambda < 1$,

$$\int_\alpha^\beta e^{i\nu x} (x-\alpha)^{\lambda-1} \phi(x) \, dx$$

$$= \sum_{n=0}^{N-1} \frac{\Gamma(n+\lambda)}{n! \, \nu^{n+\lambda}} e^{i\nu\alpha + \frac{1}{2}\lambda\pi i + \frac{1}{2}n\pi i} \phi^{(n)}(\alpha) + O\left(\frac{1}{\nu^N}\right) \qquad (11.5)$$

as $\nu \to +\infty$.

In the notation of (11.2),

$$\int_\alpha^\beta \phi(x) f(x) \, dx = \left[\sum_{n=0}^{N-1} (-1)^n \phi^{(n)}(x) I^{n+1} f(x) \right]_\alpha^\beta$$

$$+ (-1)^N \int_\alpha^\beta \phi^{(N)}(x) I^N f(x) \, dx$$

$$= \sum_{n=0}^{N-1} \frac{\Gamma(n+\lambda)}{n! \, \nu^{n+\lambda}} e^{i\nu\alpha + \frac{1}{2}\lambda\pi i + \frac{1}{2}n\pi i} \phi^{(n)}(\alpha) + R_N,$$

where

$$|R_N| = \left| \int_\alpha^\beta \phi^{(N)}(x) I^N f(x) \, dx \right|$$

$$\leqslant \frac{1}{\nu^N} \int_\alpha^\beta |\phi^{(N)}(x)| (x-\alpha)^{\lambda-1} \, dx$$

$$= O\left(\frac{1}{\nu^N}\right),$$

which proves the result. The conditions under which it was proved look rather artificial in that $\phi(x)$ and its first $N-1$ derivatives were required to vanish when $x = \beta$, a restriction which we remove in the next section. To do so, we need also the following asymptotic formula applicable when there is a singularity at $x = \beta$.

Let $\phi(x)$ be N times continuously differentiable in $\alpha \leqslant x \leqslant \beta$. Let $\phi(x)$ and its first $N-1$ derivatives vanish when $x = \alpha$. Then, if $0 < \mu < 1$,

$$\int_\alpha^\beta e^{i\nu x}(\beta - x)^{\mu-1}\phi(x)\,dx$$

$$= \sum_{n=0}^{N-1} \frac{\Gamma(n+\mu)}{n!\,\nu^{n+\mu}}\,e^{i\nu\beta - \frac{1}{2}\mu\pi i + \frac{1}{2}n\pi i}\,\phi^{(n)}(\beta) + O\!\left(\frac{1}{\nu^N}\right) \quad (11.6)$$

as $\nu \to +\infty$.

This result follows from (11.5) by a change of variable. It should be noted that the order term is not as good as that in (10.1), because Riemann's lemma is no longer applicable.

12. Use of van der Corput's neutralizer

The results of § 11 are special cases of the following theorem.

Let $\phi(x)$ be N times continuously differentiable in $\alpha \leqslant x \leqslant \beta$. Let $0 < \lambda < 1$, $0 < \mu < 1$. Then, as $\nu \to +\infty$,

$$\int_\alpha^\beta e^{i\nu x}(x-\alpha)^{\lambda-1}(\beta-x)^{\mu-1}\phi(x)\,dx$$

$$= \sum_{n=0}^{N-1} \frac{\Gamma(n+\lambda)}{n!\,\nu^{n+\lambda}}\,e^{i\nu\alpha + \frac{1}{2}\lambda\pi i + \frac{1}{2}n\pi i}\,\frac{d^n}{d\alpha^n}\{(\beta-\alpha)^{\mu-1}\phi(\alpha)\}$$

$$+ \sum_{n=0}^{N-1} \frac{\Gamma(n+\mu)}{n!\,\nu^{n+\mu}}\,e^{i\nu\beta - \frac{1}{2}\mu\pi i + \frac{1}{2}n\pi i}\,\frac{d^n}{d\beta^n}\{(\beta-\alpha)^{\lambda-1}\phi(\beta)\} + O\!\left(\frac{1}{\nu^N}\right).$$

$$(12.1)$$

And, conversely, this theorem can be derived from the results of § 11 by introducing a function which van der Corput [8] calls *a neutralizer*.

The idea of a neutralizer is a simple one. Let $f(x)$ possess continuous derivatives of all orders, and let

$$\nu_0(x) = e^{-1/x} \quad (x > 0), \qquad \nu_0(0) = 0. \qquad (12.2)$$

Then $\nu_0(x)f(x)$ possesses continuous derivatives of all orders when $x \geqslant 0$ and all these derivatives vanish when $x = 0$: such a function $\nu_0(x)$ is called a neutralizer. In this problem we need a neutralizer $\nu(x)$ with the following properties:

(i) $\nu(x)$ possesses continuous derivatives of all orders in $\alpha \leqslant x \leqslant \beta$;

(ii) $\nu(\alpha) = 1, \nu(\beta) = 0$;

(iii) the derivatives of $\nu(x)$ of all orders vanish when $x = \alpha$ and when $x = \beta$.

This function $\nu(x)$ has, of course, no connexion with the large parameter ν which occurs in the integral.

The obvious generalization of the function $\nu_0(x)$ of (12.2) is

$$\nu_1(x) = \exp\left\{ -\frac{1}{x-\alpha} - \frac{1}{\beta-x} \right\},$$

which satisfies all the conditions except (ii); for as $x \to \alpha + 0$ or $x \to \beta - 0$, $\nu_1(x) \to 0$. But

$$\nu(x) = C \int_x^\beta \nu_1(t)\,dt \quad (\alpha \leqslant x \leqslant \beta),$$

with an appropriate choice of the constant C, does satisfy the conditions. The actual form of the neutralizer chosen is not relevant in what follows.

With such a neutralizer, we write

$$\int_\alpha^\beta e^{i\nu x}(x-\alpha)^{\lambda-1}(\beta-x)^{\mu-1}\phi(x)\,dx$$

$$= \int_\alpha^\beta e^{i\nu x}(x-\alpha)^{\lambda-1}\{\nu(x)(\beta-x)^{\mu-1}\phi(x)\}\,dx$$

$$+ \int_\alpha^\beta e^{i\nu x}(\beta-x)^{\mu-1}\{[1-\nu(x)](x-\alpha)^{\lambda-1}\phi(x)\}\,dx.$$

The first of these integrals is of the type considered in (11.5) since

$$\Phi(x) = \nu(x)(\beta-x)^{\mu-1}\phi(x),$$

and all its derivatives vanish at $x = \beta$. Moreover, if we regard $\Phi(x)$ as the product of $\nu(x)$ and $(\beta-x)^{\mu-1}\phi(x)$ and differentiate

c

it n times by Leibniz's theorem, all the derivatives of $\nu(x)$ vanish at $x = \alpha$, and so

$$\Phi^{(n)}(\alpha) = \frac{d^n}{d\alpha^n}\{(\beta - \alpha)^{\mu-1}\phi(\alpha)\}.$$

The application of (11.5) to the first integral thus leads to the first sum in (12.1). Similarly, the second integral gives the second sum, which proves the result.

CHAPTER 4

THE METHOD OF STATIONARY PHASE

13. Kelvin's hydrodynamical problem

The principle of stationary phase asserts that, as $\nu \to +\infty$, the dominant terms in the asymptotic expansion of the integral

$$\int_a^b e^{i\nu f(x)} \phi(x)\, dx, \tag{13.1}$$

where $f(x)$ is real, arise from the immediate neighbourhoods of the end-points of the interval, and from the immediate neighbourhoods of the points at which the phase $\nu f(x)$ is stationary. Whilst it is difficult to formulate the principle more precisely, the underlying physical idea of Kelvin [16] is quite simple.

Let us consider the two-dimensional problem of water waves in a straight canal, a problem discussed, for example, in Lamb [17]. If we take the origin to be in the undisturbed free surface and Oy vertically upwards, the velocity potential ϕ is

$$\phi(x, y, t) = \frac{g}{\pi} \int_0^\infty \frac{\sin st}{s} e^{my} \int_{-\infty}^\infty f(a) \cos m(x-a)\, da\, dm, \tag{13.2}$$

it being assumed that the water is initially at rest with wave profile $\eta = f(x)$. Here s is a known function of m. If the canal is very deep, $s^2 = gm$; but if the canal is of depth h, then

$$s^2 = gm \tanh mh.$$

In either case the wave velocity s/m is different from the group velocity. It was the use of the principle of interference in the treatment of group velocity by Stokes and Rayleigh that suggested the principle of stationary phase.

The hydrodynamical problem is to determine the wave profile $\eta = f(x, t)$ at any subsequent time t, and this is given by

$$\eta = \frac{1}{g} \frac{\partial \phi_0}{\partial t},$$

where $\phi_0 = \phi(x, 0, t)$. If the initial elevation is confined to the immediate neighbourhood of the origin so that $f(x, 0)$ is Dirac's delta function,

$$\phi_0 = \frac{g}{\pi} \int_0^\infty \frac{\sin st}{s} \cos mx\, dm. \tag{13.3}$$

In the case of deep water ($s^2 = gm$), this integral can be transformed into what is essentially a Fresnel integral

$$\phi_0 = \frac{\sqrt{g}}{\pi \sqrt{x}} \int_{-\omega}^{\omega} \sin(\omega^2 - v^2)\, dv, \tag{13.4}$$

where $\omega^2 = \frac{1}{4}gt^2/x$. There is no corresponding simple result in more general cases, and then Kelvin's approximation becomes of great value. Nevertheless, for simplicity in presentation, we restrict ourselves to the case of deep water.

Since

$$\phi_0 = \frac{g}{2\pi} \int_0^\infty \{\sin(st + mx) + \sin(st - mx)\} \frac{dm}{s}, \tag{13.5}$$

ϕ_0 is the resultant of two superimposed systems of waves, moving to the right and the left with velocity s/m. The time variable t is positive; by symmetry, we may suppose that x is positive.

Let μ be any fixed value of m. Then if $m = \mu + u$, where u is small, the phase of the first system of waves is

$$st + mx = t\sqrt{(g\mu)} + \mu x + u\{x + \tfrac{1}{2}t\sqrt{(g/\mu)}\}$$

approximately. Kelvin's argument was that, over the small interval $(\mu - \epsilon, \mu + \epsilon)$ of values of the wave-number m, these waves are out of phase and largely cancel each other out by interference. Similarly, the waves of the second system in general largely cancel out by interference. But if $\mu = \frac{1}{4}gt^2/x^2$, the waves corresponding to the small interval $(\mu - \epsilon, \mu + \epsilon)$ of wave-numbers of the second system are in phase (at any rate correct to the first order in ϵ) and reinforce each other. This value μ of m is the value which makes the phase stationary. Kelvin's argument thus leads to the approximation

$$\phi_0 = \frac{\sqrt{g}}{2\pi \sqrt{\mu}} \int_{\mu-\epsilon}^{\mu+\epsilon} \sin(st - mx)\, dm, \tag{13.6}$$

since s varies very little near the point μ of stationary phase.

If we make the change of variable

$$m = \mu + u, \quad \mu = \tfrac{1}{4}gt^2/x^2,$$

we find that

$$st - mx = \frac{gt^2}{4x} - \frac{x^3 u^2}{gt^2},$$

correct to the second order in u. Hence approximately

$$\phi_0 = \frac{x}{\pi t} \int_{-\epsilon}^{\epsilon} \sin\left(\frac{gt^2}{4x} - \frac{x^3 u^2}{gt^2}\right) du.$$

A repetition of the same argument suggests that we may replace the limits by $\pm \infty$. If we now put

$$x^3 u^2 = gt^2 v^2,$$

we obtain

$$\phi_0 = \frac{\sqrt{g}}{\pi \sqrt{x}} \int_{-\infty}^{\infty} \sin\left(\frac{gt^2}{4x} - v^2\right) dv \qquad (13.7)$$

$$= \frac{\sqrt{g}}{\sqrt{(\pi x)}} \sin\left(\frac{gt^2}{4x} - \frac{\pi}{4}\right).$$

A comparison with the exact result (13.4) shows that Kelvin's approximation is valid when $\tfrac{1}{4}gt^2/x$ is large.

Whilst this discussion is simple physically, it does not seem to be obvious from it why the large parameter should be t^2/x. But by a change of variable, we can write (13.5) in the form

$$\phi_0 = \frac{gt}{2\pi x} \int_0^{\infty} \{\sin \nu(\theta + \sqrt{\theta}) - \sin \nu(\theta - \sqrt{\theta})\} \frac{d\theta}{\sqrt{\theta}},$$

where $\nu = gt^2/x$, which is the difference between the imaginary parts of two integrals of the form (13.1)

$$\int_a^b e^{i\nu f(\theta)} \phi(\theta) \, d\theta,$$

with a large parameter ν.

14. The principle of stationary phase

In many of the applications of the principle of stationary phase, the phase $\nu f(x)$ and the amplitude $\phi(x)$ are analytic functions of a complex variable, and we restrict our discussion to this case. Proofs under more general conditions have been given by Watson [29] and by van der Corput [7]. More recently, Erdélyi [13] has shown how some of the results can be obtained by a fairly simple method consisting of a change of variable followed by repeated integration by parts.

We shall consider the integral

$$F(\nu) = \int_a^b e^{i\nu f(x)} \phi(x)\,dx, \qquad (14.1)$$

where ν is large and positive and a and b are finite. We assume that:

(i) $f(z)$ and $\phi(z)$ are analytic functions of the complex variable z, regular in a simply connected open region D containing the segment $a \leqslant x \leqslant b$ of the real axis; and that

(ii) $f(z)$ is real on the real axis.

Since $f(z)$ is regular in D, so also is $f'(z)$. Hence $f'(z)$ has only a finite number of zeros on any compact set contained in D. In particular, $f'(x)$ has only a finite number of zeros in $a \leqslant x \leqslant b$, that is, $f(x)$ has only a finite number of stationary points there.

We can therefore divide $a \leqslant x \leqslant b$ into a finite number of closed subintervals in each of which either $f'(x)$ does not vanish, or $f'(x)$ vanishes only at the left-hand end-point, or $f'(x)$ vanishes only at the right-hand end-point. We consider these three cases separately.

(i) *If $f(x)$ has no stationary point in $\alpha \leqslant x \leqslant \beta$,*

$$I = \int_\alpha^\beta e^{i\nu f(x)} \phi(x)\,dx = \frac{\phi(\beta)}{i\nu f'(\beta)} e^{i\nu f(\beta)} - \frac{\phi(\alpha)}{i\nu f'(\alpha)} e^{i\nu f(\alpha)} + O\!\left(\frac{1}{\nu^2}\right) \quad (14.2)$$

as $\nu \to +\infty$.

Since $f'(x)$ does not vanish on $\alpha \leqslant x \leqslant \beta$, $1/f'(z)$ is an analytic function regular in an open region containing the interval. Integration by parts gives

$$I = \frac{1}{i\nu} \int_\alpha^\beta \frac{\phi}{f'} \frac{d}{dx} (e^{i\nu f})\,dx$$

$$= \frac{\phi(\beta)}{i\nu f'(\beta)} e^{i\nu f(\beta)} - \frac{\phi(\alpha)}{i\nu f'(\alpha)} e^{i\nu f(\alpha)} - \frac{1}{i\nu} \int_\alpha^\beta e^{i\nu f}\,\psi\,dx,$$

where
$$\psi(x) = \frac{d}{dx}\left(\frac{\phi}{f'}\right).$$

I is evidently $O(1/\nu)$. As ψ satisfies the same conditions as ϕ,

$$\int_\alpha^\beta e^{i\nu f}\psi\,dx$$

is also $O(1/\nu)$, and the required result follows.

(ii) *If $f(x)$ has one stationary point in $\alpha \leqslant x \leqslant \beta$, namely at $x = \alpha$, and if $f''(\alpha) > 0$,*

$$I = \int_\alpha^\beta e^{i\nu f(x)}\phi(x)\,dx = \left\{\frac{\pi}{2\nu f''(\alpha)}\right\}^{\frac{1}{2}}\phi(\alpha)\,e^{i\nu f(\alpha)+\frac{1}{4}\pi i}+O\!\left(\frac{1}{\nu}\right), \quad (14.3)$$

as $\nu \to +\infty$; but if $f''(\alpha) < 0$,

$$I = \int_\alpha^\beta e^{i\nu f(x)}\phi(x)\,dx = \left\{\frac{\pi}{-2\nu f''(\alpha)}\right\}^{\frac{1}{2}}\phi(\alpha)\,e^{i\nu f(\alpha)-\frac{1}{4}\pi i}+O\!\left(\frac{1}{\nu}\right). \quad (14.4)$$

This shows that, if $\phi(\alpha) \neq 0$, I is $O(1/\sqrt{\nu})$ and so is more important than the contribution of an interval containing no stationary point.

By (i), we can take β as near to α as we please. We choose β so that it lies inside a circle with centre α within which all the analytic functions which occur are regular. Moreover, since $f'(\alpha) = 0$, $f''(\alpha) > 0$, we can also choose β so that $f(x)$ increases steadily as x increases from α to β. If we introduce a new variable u defined by
$$f(x) = f(\alpha)+u^2,$$

where u is positive when $\alpha < x \leqslant \beta$, we obtain

$$I = e^{i\nu f(\alpha)}\int_0^\epsilon e^{i\nu u^2}\phi(x)\frac{dx}{du}\,du,$$

where $\epsilon = \sqrt{\{f(\beta)-f(\alpha)\}}$.

Now consider the equation
$$f(z) = f(\alpha)+w^2.$$

Since $f'(\alpha) = 0$, this equation has two solutions

$$z = \alpha+\sum_1^\infty b_n w^n$$

and
$$z = \alpha+\sum_1^\infty b_n(-w)^n,$$

both regular in a neighbourhood of $w = 0$, the first coefficient being

$$b_1 = \sqrt{\left(\frac{2}{f''(\alpha)}\right)}.$$

If we take the positive value of b_1, the relation between the variables x and u is

$$x = \alpha + \sum_1^\infty b_n u^n.$$

Now

$$\int_\alpha^z \phi(t)\, dt$$

is an analytic function of z, regular in a neighbourhood of α, where

$$z = \alpha + \sum_1^\infty b_n w^n$$

is regular in a neighbourhood of $w = 0$. Hence this integral is itself an analytic function of w, which vanishes when $w = 0$, and is expansible as a convergent power series

$$\sum_0^\infty c_n w^{n+1}/(n+1).$$

It follows that

$$\phi(z)\frac{dz}{dw} = \sum_1^\infty c_n w^n,$$

where

$$c_0 = \phi(\alpha)\sqrt{\left(\frac{2}{f''(\alpha)}\right)}$$

We may therefore write

$$\phi(z)\frac{dz}{dw} = c_0 + w\psi(w),$$

where $\psi(w)$ is itself regular in a neighbourhood of $w = 0$, and so

$$I = e^{i\nu f(\alpha)}\int_0^\epsilon e^{i\nu u^2}\{c_0 + u\psi(u)\}\, du$$

$$= \{I_1 + I_2\}\, e^{i\nu f(\alpha)},$$

say. Here

$$I_1 = \int_0^\epsilon c_0\, e^{i\nu u^2}\, du = \frac{c_0}{2\sqrt{\nu}}\int_0^{\nu\epsilon^2}\frac{e^{i\nu}}{\sqrt{\nu}}\, dv$$

$$= \frac{c_0}{2\sqrt{\nu}}\int_0^\infty \frac{e^{iv}}{\sqrt{v}}\, dv - \frac{c_0}{2\sqrt{\nu}}\int_{\nu\epsilon^2}^\infty \frac{e^{iv}}{\sqrt{v}}\, dv$$

$$= \tfrac{1}{2}c_0\left(\frac{\pi}{\nu}\right)^{\frac{1}{2}} e^{\frac{1}{4}\pi i} + O\left(\frac{1}{\nu}\right),$$

the order term being obtained by integration by parts; and

$$|I_2| = \left| \int_0^\epsilon e^{i\nu u^2} u\psi(u)\, du \right|$$

$$= \left| \frac{1}{2i\nu} \left\{ e^{i\nu\epsilon^2} \psi(\epsilon) - \psi(0) - \int_0^\epsilon e^{i\nu u^2} \psi'(u)\, du \right\} \right|$$

$$\leqslant \frac{1}{2\nu} \left\{ |\psi(\epsilon)| + |\psi(0)| + \int_0^\epsilon |\psi'(u)|\, du \right\} = O\left(\frac{1}{\nu}\right).$$

We have thus proved that, as $\nu \to +\infty$,

$$\int_\alpha^\beta e^{i\nu f(x)} \phi(x)\, dx = \tfrac{1}{2} c_0 \left(\frac{\pi}{\nu}\right)^{\frac{1}{2}} e^{i\nu f(\alpha) + \frac{1}{4}\pi i} + O\left(\frac{1}{\nu}\right),$$

which is the result stated. The proof when $f''(\alpha)$ is negative follows the same lines.

(iii) *If $f(x)$ has one stationary point in $\alpha \leqslant x \leqslant \beta$, namely at $x = \beta$, and if $f''(\beta) > 0$,*

$$I = \int_\alpha^\beta e^{i\nu f(x)} \phi(x)\, dx = \left\{ \frac{\pi}{2\nu f''(\beta)} \right\}^{\frac{1}{2}} \phi(\beta)\, e^{i\nu f(\beta) + \frac{1}{4}\pi i} + O\left(\frac{1}{\nu}\right) \quad (14.5)$$

as $\nu \to +\infty$; but if $f''(\beta) < 0$,

$$I = \int_\alpha^\beta e^{i\nu f(x)} \phi(x)\, dx = \left\{ \frac{\pi}{-2\nu f''(\beta)} \right\}^{\frac{1}{2}} \phi(\beta)\, e^{i\nu f(\beta) - \frac{1}{4}\pi i} + O\left(\frac{1}{\nu}\right). \quad (14.6)$$

This can be proved in the same way.

If, in (ii), the first derivative which does not vanish at $x = \alpha$ is the nth, the integral is of order $O(1/\nu^{1/n})$. A result for $n = 3$ is

(iv) *If $f(x)$ has one stationary point in $\alpha \leqslant x \leqslant \beta$, namely at $x = \alpha$, and if $f''(\alpha) = 0, f'''(\alpha) > 0$,*

$$I = \int_\alpha^\beta e^{i\nu f(x)} \phi(x)\, dx = \Gamma(\tfrac{4}{3}) \left\{ \frac{6}{\nu f'''(\alpha)} \right\}^{\frac{1}{3}} e^{i\nu f(\alpha) + \frac{1}{6}\pi i} \phi(\alpha) + O\left(\frac{1}{\nu^{\frac{2}{3}}}\right)$$

$$(14.7)$$

as $\nu \to +\infty$.

One could proceed further and consider more complicated stationary points. But in every case it would turn out that the dominant part of the integral arises from the neighbourhood of the stationary point. Erdélyi [13], using real variable methods, has discussed the case

$$f'(x) = (x-\alpha)^{\rho-1}(\beta-x)^{\sigma-1}f_1(x) \quad (\rho, \sigma \geqslant 1),$$

where $f_1(x)$ is positive and N times continuously differentiable,

$$\phi(x) = (x-\alpha)^{\lambda-1}(\beta-x)^{\mu-1}\phi_1(x) \quad (0 < \lambda, \mu \leqslant 1),$$

and $\phi_1(x)$ is N times continuously differentiable. He uses integration by parts and the van der Corput neutralizer to enable him to consider the points α and β separately. If $f_1(x)$ and $\phi_1(x)$ possess continuous derivatives of all orders, this process gives a complete asymptotic expansion, not merely the dominant term.

15. Applications to Bessel functions

The principle of stationary phase was applied by Nicholson [20] and Rayleigh [24] to obtain approximate formulae for the Bessel functions $J_\nu(\nu)$ and $J_\nu(\nu \sec \beta)$ when ν is large and positive and β is a positive acute angle.

In the formula

$$J_\nu(\nu \sec \beta) = \frac{1}{\pi} \int_0^\pi \cos \nu(\theta - \sec \beta \sin \theta)\, d\theta$$
$$- \frac{\sin \nu\pi}{\pi} \int_0^\infty e^{-\nu(t+\sec \beta \sinh t)}\, dt,$$

the second term is $O(1/\nu)$; for

$$0 < \int_0^\infty e^{-\nu(t+\sec \beta \sinh t)}\, dt < \int_0^\infty e^{-\nu t} = \frac{1}{\nu}.$$

The first term is the real part of

$$I = \frac{1}{\pi} \int_0^\pi e^{i\nu f(\theta)}\, d\theta,$$

where $f(\theta) = \theta - \sec \beta \sin \theta.$

The function $f(\theta)$ is stationary when $\theta = \beta$, and $f''(\beta) = \tan \beta$. We may therefore apply (14.3) to the interval $\beta \leqslant \theta \leqslant \pi$, and (14.5) to $0 \leqslant \theta \leqslant \beta$. It follows that

$$I = \left\{\frac{2}{\pi\nu \tan \beta}\right\}^{\frac{1}{2}} e^{i\nu(\beta - \tan \beta) + \frac{1}{4}\pi i} + O\left(\frac{1}{\nu}\right)$$

and hence that

$$J_\nu(\nu \sec \beta) = \left\{\frac{2}{\pi\nu \tan \beta}\right\}^{\frac{1}{2}} \cos \{\nu(\beta - \tan \beta) + \tfrac{1}{4}\pi\} + O\left(\frac{1}{\nu}\right),$$

as $\nu \to +\infty$.

Similarly, $\qquad\qquad J_\nu(\nu) = \mathscr{R}I + O(1/\nu),$

where $\qquad\qquad\qquad I = \frac{1}{\pi}\int_0^\pi e^{i\nu f(\theta)}\,d\theta$

and $\qquad\qquad\qquad f(\theta) = \theta - \sin \theta.$

The phase is stationary only at $\theta = 0$; and $f''(0) = 0$, $f'''(0) = 1$. By (14.7), we have

$$I = \frac{1}{\pi}\,\Gamma(\tfrac{4}{3})\left(\frac{6}{\nu}\right)^{\frac{1}{3}} e^{\frac{1}{6}\pi i} + O\left(\frac{1}{\nu^{\frac{2}{3}}}\right),$$

and so $\qquad J_\nu(\nu) = \frac{1}{\pi}\,\Gamma(\tfrac{4}{3})\left(\frac{6}{\nu}\right)^{\frac{1}{3}}\frac{3^{\frac{1}{2}}}{2} + O\left(\frac{1}{\nu^{\frac{2}{3}}}\right),$

or $\qquad\qquad J_\nu(\nu) = \frac{\Gamma(\tfrac{1}{3})}{2^{\frac{2}{3}}3^{\frac{1}{6}}\pi\nu^{\frac{1}{3}}} + O\left(\frac{1}{\nu^{\frac{2}{3}}}\right).$

The order term is, in fact, a rather poor approximation. By means of what he described as 'some tedious integrations by parts', Watson obtained a better error term $O(1/\nu^{\frac{4}{3}})$.

16. Multiple integrals

The method of stationary phase can also be applied to double integrals of the form

$$\iint e^{i\nu f(x,\,y)}\,\phi(x, y)\,dx\,dy,$$

which arise in the theory of diffraction. An account of this and a bibliography of other work in this field have been given by Jones and Klein [15].

CHAPTER 5

THE METHOD OF LAPLACE

17. The Laplace approximation

In the integral

$$f(\nu) = \int_{\alpha}^{\beta} \phi(x) \, e^{\nu h(x)} \, dx, \qquad (17.1)$$

let ν be a large positive constant, and let $\phi(x)$ and $h(x)$ be real and continuous in $\alpha \leqslant x \leqslant \beta$. Then the major contributions to the value of the integral evidently arise from the neighbourhoods of the points at which $h(x)$ attains its greatest value. This greatest value may be a maximum in the calculus sense or a supremum; and the Laplace approximation is different in the two cases.

We assume that the derivatives $h'(x)$ and $h''(x)$ are continuous. Let us consider first the case when $h(x)$ attains its greatest value, a supremum, at $x = \alpha$ at which $h'(x)$ is negative. Then we can find an interval $\alpha \leqslant x \leqslant \alpha + \eta$ in which $h'(x)$ is negative and bounded from zero. The new variable t defined by

$$h(x) = h(\alpha) - t$$

increases steadily from 0 to τ, say, as x increases from α to $\alpha + \eta$. The dominant part of $f(\nu)$, when ν is large and positive, is given by

$$f(\nu) \sim \int_{\alpha}^{\alpha+\eta} \phi(x) \, e^{\nu h(x)} \, dx = -\int_{0}^{\tau} \frac{\phi(x)}{h'(x)} \, e^{\nu h(\alpha) - \nu t} \, dt.$$

For sufficiently small η, we may replace $\phi(x)$ and $h'(x)$ by $\phi(\alpha)$ and $h'(\alpha)$ and obtain

$$f(\nu) \sim -\frac{\phi(\alpha)}{h'(\alpha)} \, e^{\nu h(\alpha)} \int_{0}^{\tau} e^{-\nu t} \, dt$$

$$= -\frac{\phi(\alpha)}{h'(\alpha)} \frac{e^{\nu h(\alpha)}}{\nu} \, (1 - e^{-\nu \tau}).$$

This crude argument leads therefore to the approximation

$$f(\nu) \sim -\frac{\phi(\alpha)}{h'(\alpha)} \frac{e^{\nu h(\alpha)}}{\nu} \qquad (17.2)$$

as $\nu \to +\infty$. If $\phi(\alpha) = 0$, a similar argument gives

$$f(\nu) = o\left(\frac{e^{\nu h(\alpha)}}{\nu}\right),$$

where the precise order depends on the behaviour of $\phi(x)$ at $x = \alpha$.

Similarly, if $h(x)$ attains its supremum at $x = \beta$ where $h'(x)$ is positive,

$$f(\nu) \sim \frac{\phi(\beta)}{h'(\beta)} \frac{e^{\nu h(\beta)}}{\nu}.$$

There is no loss of generality in assuming that the supremum is attained at an end-point, since the range of integration can always be broken up into subintervals each of which has that property.

If $h(x)$ has a finite number of maxima in the calculus sense, we can again break up the range of integration into a finite number of subintervals, in each of which $h(x)$ attains a maximum in the calculus sense at one end but not at the other. We may restrict our attention, therefore, to the case of the integral (17.1) when $h(x)$ has a maximum in the calculus sense at $x = \alpha$ and $h(x) < h(\alpha)$ when $\alpha < x \leqslant \beta$. A similar argument will apply when the maximum is attained at $x = \beta$.

In the simplest case, $h'(\alpha) = 0$ and $h''(\alpha) < 0$. We can then find an interval $\alpha \leqslant x \leqslant \alpha + \eta$ in which $h''(x)$ is negative and bounded from zero and $h'(x)$ is negative. Following Laplace [19] we define the new variable t by

$$h(x) = h(\alpha) - t^2,$$

where t increases steadily from 0 to τ, say, as x increases from α to $\alpha + \eta$. The dominant part of $f(\nu)$, when ν is large and positive, is then

$$f(\nu) \sim \int_{\alpha}^{\alpha+\eta} \phi(x) e^{\nu h(x)} dx$$

$$= -\int_0^\tau \phi(x) e^{\nu h(x) - \nu t^2} \frac{2t}{h'(x)} dt.$$

For sufficiently small η, we may replace $\phi(x)$ by $\phi(\alpha)$. But, by the mean-value theorem,

$$t^2 = h(\alpha) - h(x) = -\tfrac{1}{2}(x-\alpha)^2 h''(\xi),$$

where $\alpha < \xi < \alpha + \eta$, and

$$h'(x) = (x - \alpha) h''(\xi_1),$$

where $\alpha < \xi_1 < \alpha + \eta$. Therefore

$$\frac{2t}{h'(x)} = \frac{\sqrt{(-2h''(\xi))}}{h''(\xi_1)},$$

which may be replaced by $-1/\sqrt{\{-\frac{1}{2}h''(\alpha)\}}$ for sufficiently small η. This leads to the Laplace approximation

$$f(\nu) \sim 2 \frac{\phi(\alpha) e^{\nu h(\alpha)}}{\sqrt{\{-2h''(\alpha)\}}} \int_0^\tau e^{-\nu t^2} dt.$$

But, by the same argument,

$$f(\nu) \sim 2 \frac{\phi(\alpha) e^{\nu h(\alpha)}}{\sqrt{\{-2h''(\alpha)\}}} \int_0^\infty e^{-\nu t^2} dt$$

since in this last integral the dominant part comes from the neighbourhood of $t = 0$ at which the integrand has its maximum. Therefore

$$f(\nu) \sim \phi(\alpha) e^{\nu h(\alpha)} \left[\frac{-\pi}{2\nu h''(\alpha)} \right]^{\frac{1}{2}}, \qquad (17.3)$$

when $\phi(\alpha) \neq 0$. But if $\phi(\alpha) = 0$,

$$f(\nu) = o\{e^{\nu h(\alpha)}/\sqrt{\nu}\},$$

where the precise order depends on the behaviour of $\phi(x)$ at $x = \alpha$. The same formulae hold, with α replaced by β, when the calculus maximum is attained at $x = \beta$. A similar argument can be used when the first $(n-1)$ derivatives of $h(x)$ vanish at $x = \alpha$ and the nth derivative is negative there, the corresponding change of variable being

$$h(x) = h(\alpha) - t^n.$$

Since it would be difficult to prove the Laplace approximation on the lines suggested above, we give a different proof, based essentially on that of Pólya and Szegö [23] under conditions sufficiently general for many applications. For a proof under less restrictive conditions, we refer the reader to Widder [32].

18. Proof of the Laplace approximation

Let $\phi(x)$ and $h(x)$ be two real continuous functions defined in the finite or semi-infinite interval $\alpha \leqslant x \leqslant \beta$, such that

(i) *$\phi(x)\,e^{\nu h(x)}$ is absolutely integrable over the interval for every positive value of ν;*

(ii) *$h(x)$ has a single maximum in the interval, namely at $x = \alpha$; and the supremum of $h(x)$ in any closed subinterval not containing α is less than $h(\alpha)$;*

(iii) *$h''(x)$ is continuous; and $h'(\alpha) = 0$, $h''(\alpha) < 0$. Then, as $\nu \to +\infty$,*

$$\int_\alpha^\beta \phi(x)\,e^{\nu h(x)}\,dx \sim \phi(\alpha)\,e^{\nu h(\alpha)} \left[\frac{-\pi}{2\nu h''(\alpha)}\right]^{\frac{1}{2}}.$$

Under the conditions stated, we can assign arbitrarily a positive number ϵ and then choose a positive number $\delta\,(<\beta-\alpha)$ such that

$$\phi(\alpha)-\epsilon \leqslant \phi(x) \leqslant \phi(\alpha)+\epsilon,$$

$$h''(\alpha)-\epsilon \leqslant h''(x) \leqslant h''(\alpha)+\epsilon < 0,$$

when $\alpha \leqslant x \leqslant \alpha+\delta$. Since in this subinterval

$$h(x) = h(\alpha) + \tfrac{1}{2}(x-\alpha)^2\,h''(\xi),$$

where $\alpha < \xi < \alpha+\delta$, $h(x)-h(\alpha)$ lies between $-\tfrac{1}{2}B(x-\alpha)^2$ and $-\tfrac{1}{2}A(x-\alpha)^2$ where A and B are the positive constants

$$A = -h''(\alpha)-\epsilon, \quad B = -h''(\alpha)+\epsilon.$$

Hence

$$\int_\alpha^{\alpha+\delta} \phi(x)\,e^{\nu h(x)}\,dx$$

lies between

$$\{\phi(\alpha)-\epsilon\}\,e^{\nu h(\alpha)} \int_\alpha^{\alpha+\delta} e^{-\frac{1}{2}\nu B(x-\alpha)^2}\,dx$$

and

$$\{\phi(\alpha)+\epsilon\}\,e^{\nu h(\alpha)} \int_\alpha^{\alpha+\delta} e^{-\frac{1}{2}\nu A(x-\alpha)^2}\,dx.$$

But since

$$\int_\alpha^{\alpha+\delta} e^{-\frac{1}{2}\nu A(x-\alpha)^2}\,dx = \int_0^\infty e^{-\frac{1}{2}\nu A u^2}\,du - \int_\delta^\infty e^{-\frac{1}{2}\nu A u^2}\,du$$

$$= \left\{\frac{\pi}{2\nu A}\right\}^{\frac{1}{2}} \{1 + O(e^{-\frac{1}{2}\nu A\delta^2})\}$$

when ν is large, it follows that

$$\int_{\alpha}^{\alpha+\delta} \phi(x)\, e^{\nu h(x)}\, dx \leqslant \{\phi(\alpha)+\epsilon\}\, e^{\nu h(\alpha)} \left\{\frac{\pi}{2\nu A}\right\}^{\frac{1}{2}} \{1+O(e^{-\frac{1}{2}\nu A\delta^2})\},$$

and similarly

$$\int_{\alpha}^{\alpha+\delta} \phi(x)\, e^{\nu h(x)}\, dx \geqslant \{\phi(\alpha)-\epsilon\}\, e^{\nu h(\alpha)} \left\{\frac{\pi}{2\nu B}\right\}^{\frac{1}{2}} \{1+O(e^{-\frac{1}{2}\nu B\delta^2})\}.$$

For the rest of the interval, we have

$$\left|\int_{\alpha+\delta}^{\beta} \phi(x)\, e^{\nu h(x)}\, dx\right| \leqslant \int_{\alpha+\delta}^{\beta} |\phi(x)|\, e^{h(x)}\, e^{(\nu-1)M}\, dx,$$

where
$$M = \sup_{\alpha+\delta \leqslant x \leqslant \beta} h(x) < h(\alpha)$$

by (ii). Hence, by (i)

$$\left|\int_{\alpha+\delta}^{\beta} \phi(x)\, e^{\nu h(x)}\, dx\right| \leqslant e^{(\nu-1)M} \int_{\alpha}^{\beta} |\phi(x)|\, e^{h(x)}\, dx = K\, e^{(\nu-1)M}.$$

We now have

$$\{\phi(\alpha)-\epsilon\}\left\{\frac{\pi}{2B}\right\}^{\frac{1}{2}} \{1+O(e^{-\frac{1}{2}\nu B\delta^2})\} - \frac{K\sqrt{\nu}}{e^M}\, e^{\nu\{M-h(\alpha)\}}$$

$$\leqslant \int_{\alpha}^{\beta} \phi(x)\, e^{\nu h(x)}\, dx\, \sqrt{\nu}\, e^{-\nu h(\alpha)}$$

$$\leqslant \{\phi(\alpha)+\epsilon\}\left\{\frac{\pi}{2A}\right\}^{\frac{1}{2}} \{1+O(e^{-\frac{1}{2}\nu A\delta^2})\} + \frac{K\sqrt{\nu}}{e^M}\, e^{\nu\{M-h(\alpha)\}}.$$

Let $\nu \to +\infty$. Then

$$\{\phi(\alpha)-\epsilon\}\left\{\frac{-\pi}{2h''(\alpha)-2\epsilon}\right\}^{\frac{1}{2}} \leqslant \overline{\lim} \int_{\alpha}^{\beta} \phi(x)\, e^{\nu h(x)}\, dx\, \sqrt{\nu}\, e^{-\nu h(\alpha)}$$

$$\leqslant \{\phi(\alpha)+\epsilon\}\left\{\frac{-\pi}{2h''(\alpha)+2\epsilon}\right\}^{\frac{1}{2}}.$$

But as ϵ is arbitrary, this implies that

$$\lim_{\nu \to \infty} \int_{\alpha}^{\beta} \phi(x)\, e^{\nu h(x)}\, dx\, \sqrt{\nu}\, e^{-\nu h(\alpha)} = \phi(\alpha)\left\{\frac{-\pi}{2h''(\alpha)}\right\}^{\frac{1}{2}},$$

that is, that

$$\int_{\alpha}^{\beta} \phi(x)\, e^{\nu h(x)}\, dx \sim \phi(\alpha)\, e^{\nu h(\alpha)}\left\{\frac{-\pi}{2\nu h''(\alpha)}\right\}^{\frac{1}{2}},$$

which was to be proved. The result when $\phi(\alpha)$ is zero also follows.

An alternative form of this approximation is obtained by putting $e^{h(x)} = f(x)$, namely, that if $f(x)$ attains its maximum at $x = \alpha$ where $f'(\alpha) = 0, f''(\alpha) < 0$, then

$$\int_\alpha^\beta \phi(x)\{f(x)\}^\nu\, dx \sim \phi(\alpha)\{f(\alpha)\}^{\nu+\frac{1}{2}} \left\{\frac{-\pi}{2\nu f''(\alpha)}\right\}^{\frac{1}{2}}$$

as $\nu \to \infty$.

19. Some examples of the Laplace approximation

A simple example of the Laplace approximation is provided by the Bessel function $I_n(t)$ of integer order n, which has the integral representation

$$I_n(t) = \frac{1}{\pi} \int_0^\pi e^{t\cos\theta} \cos n\theta\, d\theta.$$

Here $\phi(\theta) = \cos n\theta$; and $h(\theta) = \cos\theta$ decreases steadily from a maximum at $\theta = 0$. Since $h(0) = 1$, $h'(0) = 0$, $h''(0) = -1$, it follows that

$$I_n(t) \sim \frac{e^t}{\sqrt{(2\pi t)}}$$

as $t \to +\infty$.

As an example of the alternative formulation, let us consider Laplace's first integral for the Legendre polynomial $P_n(\mu)$, namely

$$P_n(\mu) = \frac{1}{\pi} \int_0^\pi \{\mu + (\mu^2-1)^{\frac{1}{2}} \cos\theta\}^n\, d\theta,$$

where $\mu > 1$ and the square root is positive. Here $\phi(\theta) = 1$ and

$$f(\theta) = \mu + (\mu^2 - 1)^{\frac{1}{2}} \cos\theta.$$

Since $f(\theta)$ has its greatest value at $\theta = 0$ where $f'(\theta) = 0$, $f''(\theta) = -(\mu^2-1)^{\frac{1}{2}}$, we get at once the result that

$$P_n(\mu) \sim \frac{1}{\sqrt{(2\pi n)}} \frac{\{\mu + (\mu^2-1)^{\frac{1}{2}}\}^{n+\frac{1}{2}}}{(\mu^2-1)^{\frac{1}{4}}}$$

as $n \to \infty$.

As a last example, we take the Gamma function defined by Euler's integral

$$\Gamma(\nu+1) = \int_0^\infty e^{-u} u^\nu\, du,$$

D

and find the asymptotic approximation when ν is large and positive. The theory as it stands is inapplicable, since u has no maximum. But if we put $u = \nu t$, we obtain

$$\Gamma(\nu + 1) = \nu^{\nu+1} \int_0^\infty e^{\nu(-t+\log t)}\,dt.$$

This integral is of the desired form with $h(t) = -t + \log t$, which has a single maximum at $t = 1$, with $h'(1) = 0$, $h''(1) = -1$. If we apply the approximation to the two intervals $0 \leqslant t \leqslant 1$ and $t \geqslant 1$, we obtain
$$\Gamma(\nu + 1) \sim (2\pi\nu)^{\frac{1}{2}}\,\nu^\nu\,e^{-\nu}$$
as $\nu \to +\infty$.

20. Generalizations of Laplace's method

In our discussion of the Laplace approximation, we considered integrals of the form
$$\int_\alpha^\beta \phi(x)\,e^{\nu h(x)}\,dx$$

as $\nu \to \infty$, where $h(x)$ has a single maximum at a point ξ in the finite or infinite interval $\alpha \leqslant x \leqslant \beta$. For convenience, we broke up the interval so that the maximum was attained at an end-point, but this was not essential. It would be possible to generalize the method to cover integrals of the form
$$\int_\alpha^\beta \phi(x, \nu)\,e^{h(x, \nu)}\,dx,$$

where $\phi(x, \nu)$ is bounded as $\nu \to \infty$ and $h(x, \nu)$ has a single maximum ξ; but this stationary point ξ would no longer be fixed—it would vary with ν.

The expression of the integrand as a product $\phi\,e^h$ is somewhat arbitrary, and different factorizations may lead to different asymptotic formulae valid in different circumstances. It is customary to change the variable, if it be possible, to make the stationary point independent of ν. It is not always possible, and it is not necessary to do this.

For example, in the case of the Gamma function,
$$\Gamma(\nu + 1) = \int_0^\infty e^{\nu \log x - x}\,dx,$$

we could not take $\phi(x) = e^{-x}$, $h(x) = \log x$, because $\log x$ has no stationary point. But we could take $\phi(x) = 1, h(x, \nu) = \nu \log x - x$; the function $h(x, \nu)$ has a single maximum at $x = \nu$, and the asymptotic approximation could be worked out on that basis. We avoided this by making the change of variable $x = \nu t$.

In the case of the Bessel function

$$K_\nu(a) = \frac{1}{2} \int_{-\infty}^{\infty} e^{\nu x - a \cosh x}\, dx,$$

where ν and a are positive and ν is large, Laplace's method is not applicable with the factorization $h(x) = x$, $\phi(x) = e^{-a \cosh x}$ since $h(x)$ has no maximum. But $\nu x - a \cosh x$ has a single maximum at $x = \sinh^{-1}(\nu/a)$, which varies with ν. If we put

$$x = \sinh^{-1}(\nu/a) + t,$$

we obtain

$$K_\nu(a) = \frac{1}{2}\left\{\frac{\nu + \sqrt{(\nu^2+a^2)}}{a}\right\}^\nu \int_{-\infty}^{\infty} e^{\nu(t-e^t)} \phi(t, \nu)\, dt,$$

where $\qquad \phi(t, \nu) = \exp[-a^2 \cosh t / \{\nu + \sqrt{(\nu^2+a^2)}\}].$

The function $t - e^t$ has a single maximum at $t = 0$, but we have obtained this simple form at the expense of introducing ν into ϕ. This is quite harmless, since, in any finite interval $-\alpha \leqslant t \leqslant \alpha$, $\phi(t, \nu)$ is continuous and

$$\phi(\alpha, \nu) \leqslant \phi(t, \nu) \leqslant \phi(0, \nu);$$

hence $\phi(t, \nu) \to 1$ as $\nu \to \infty$, uniformly in t. And $\phi(t, \nu) \leqslant 1$ for all t, ν.

We now have

$$\int_{-\alpha}^{\alpha} e^{\nu(t-e^t)} \phi(t, \nu)\, dt = \phi(t_0, \nu) \int_{-\alpha}^{\alpha} e^{\nu(t-e^t)}\, dt,$$

where $-\alpha \leqslant t_0 \leqslant \alpha$, and so

$$\int_{-\alpha}^{\alpha} e^{\nu(t-e^t)} \phi(t, \nu)\, dt \sim \int_{-\infty}^{\infty} e^{\nu(t-e^t)}\, dt$$

$$\sim e^{-\nu} \int_{-\infty}^{\infty} e^{-\frac{1}{2}\nu t^2}\, dt$$

$$= e^{-\nu}\sqrt{\left(\frac{2\pi}{\nu}\right)}.$$

To prove that the contributions of the intervals $t \geqslant \alpha$ and $t \leqslant -\alpha$ are small, we write $t = \alpha + \tau$ and recall that $\phi(t, \nu) \leqslant 1$ for all t, ν. Then, using the inequality $e^\tau \geqslant 1 + \tau$ $(\tau \geqslant 0)$, we have

$$0 \leqslant \int_\alpha^\infty e^{\nu(t-e^t)} \phi(t, \nu)\, dt \leqslant e^{\nu(\alpha-e^\alpha)} \int_0^\infty e^{\nu\tau - \nu e^\alpha(e^\tau - 1)}\, d\tau$$

$$\leqslant e^{\nu(\alpha-e^\alpha)} \int_0^\infty e^{-\nu\tau(e^\alpha - 1)}\, d\tau$$

$$= \frac{e^{-\nu(e^\alpha - \alpha)}}{\nu(e^\alpha - 1)} = o\!\left(\frac{e^{-\nu}}{\nu}\right),$$

since $e^\alpha - \alpha > 1$. Similarly for $t \leqslant -\alpha$.

We have thus shown that, when a is positive and $\nu \to +\infty$,

$$K_\nu(a) \sim \frac{\{\nu + \sqrt{(\nu^2 + a^2)}\}^\nu}{2a^\nu} e^{-\nu} \sqrt{\left(\frac{2\pi}{\nu}\right)},$$

or, more conveniently,

$$K_\nu(a) \sim \frac{2^\nu \nu^\nu e^{-\nu}}{a^\nu} \sqrt{\left(\frac{\pi}{2\nu}\right)},$$

a result which could have been obtained when ν is not an integer from the formula

$$K_\nu(a) = \frac{\pi}{2 \sin \nu\pi} \{I_{-\nu}(a) - I_\nu(a)\}.$$

A more difficult example is provided by the parabolic cylinder function defined by

$$D_{-\nu-1}(a) = \frac{e^{-\frac{1}{4}a^2}}{\Gamma(\nu+1)} \int_0^\infty e^{-ax - \frac{1}{2}x^2} x^\nu\, dx,$$

when $\mathscr{R}\nu > -1$. We propose to find an asymptotic approximation, due to Cherry [4], valid when a is positive and $\nu \to +\infty$. If we put $x = s\sqrt{\nu}$, we get

$$D_{-\nu-1}(a) = \frac{e^{-\frac{1}{4}a^2}}{\Gamma(\nu+1)} \nu^{\frac{1}{2}\nu + \frac{1}{2}} \int_0^\infty e^{h(s, \nu)}\, ds,$$

where $\qquad h(s, \nu) = \nu \log s - \tfrac{1}{2}\nu s^2 - as\sqrt{\nu}.$

The function $h(s, \nu)$ has a single maximum in the range of integration at

$$s = \left(1 + \frac{a^2}{4\nu}\right)^{\frac{1}{2}} - \frac{a}{2\sqrt{\nu}},$$

which is approximately 1 when ν is large. Writing $s = 1 + t$, we find that

$$D_{-\nu-1}(a) = \frac{e^{-\frac{1}{4}a^2 - a\sqrt{\nu}}\,\nu^{\frac{1}{2}\nu + \frac{1}{2}}}{\Gamma(\nu+1)}\,I,$$

where $\quad I = \displaystyle\int_{-1}^{\infty} \exp\{\nu\log(1+t) - \tfrac{1}{2}\nu(1+t)^2\}\,e^{-at\sqrt{\nu}}\,dt.$

This is quite different from the previous problem in that $e^{-at\sqrt{\nu}}$ does not tend to a continuous limit as $\nu \to +\infty$.

To avoid continually writing $\sqrt{\nu}$, it is convenient to write $\sqrt{\nu} = \kappa$. Putting $t = u/\kappa$, we have

$$I = \int_{-\kappa}^{\infty} \exp\left\{\kappa^2\log\left(1 + \frac{u}{\kappa}\right) - \tfrac{1}{2}(u+\kappa)^2 - au\right\}\frac{du}{\kappa}.$$

Now by the mean value theorem

$$\log\left(1 + \frac{u}{\kappa}\right) = \frac{u}{\kappa} - \frac{u^2}{2\kappa^2} + \frac{u^3}{3(\kappa + u_1)^3},$$

where u_1 lies between 0 and u. Hence

$$\kappa^2\log\left(1 + \frac{u}{\kappa}\right) - \tfrac{1}{2}(u+\kappa)^2 - au = -\tfrac{1}{2}\kappa^2 - u^2 - au + \frac{\kappa^2 u^3}{3(\kappa + u_1)^3}.$$

It follows that $\quad I = \dfrac{e^{-\frac{1}{2}\kappa^2}}{\kappa}\displaystyle\int_{-\kappa}^{\infty} e^{-u^2 - au}\psi(u, \kappa)\,du,$

where $\qquad\qquad \psi(u, \kappa) = \exp\left\{\dfrac{\kappa^2 u^3}{3(\kappa + u_1)^3}\right\}.$

When $\kappa > 1$ and $0 < \epsilon < 1$, the interval $-\kappa^\epsilon \leqslant u \leqslant \kappa^\epsilon$ lies inside the range of integration, and in this interval

$$\left|\frac{\kappa^2 u^3}{3(\kappa + u_1)^3}\right| \leqslant \frac{\kappa^{2+3\epsilon}}{3(\kappa - \kappa^\epsilon)^3} = \frac{\kappa^{3\epsilon-1}}{3(1 - \kappa^{\epsilon-1})^3} < \kappa^{3\epsilon-1}$$

for all sufficiently large values of κ. Hence $\psi(u, \kappa)$ lies between $\exp\{\pm\kappa^{3\epsilon-1}\}$. If now we choose ϵ so that $0 < \epsilon < \tfrac{1}{3}$, $\psi(u, \kappa)$ tends to 1 as $\kappa \to \infty$, uniformly with respect to u. Writing η for κ^ϵ, we see that, as $\kappa \to \infty$,

$$\frac{e^{-\frac{1}{2}\kappa^2}}{\kappa}\int_{-\eta}^{\eta} e^{-u^2 - a\kappa}\psi(u, \kappa)\,du \sim \frac{e^{-\frac{1}{2}\kappa^2}}{\kappa}\int_{-\infty}^{\infty} e^{-u^2 - au}\,du$$

$$= e^{-\frac{1}{2}\nu + \frac{1}{4}a^2}\Big/\left(\sqrt{\frac{\pi}{\nu}}\right).$$

The contribution to I of the interval $u \geqslant \eta$ is positive and is equal to

$$\frac{1}{\kappa} \int_\eta^\infty \exp\left\{\kappa^2 \log\left(1+\frac{u}{\kappa}\right) - \tfrac{1}{2}(\kappa+u)^2 - au\right\} du$$

$$\leqslant \frac{1}{\kappa} \int_\eta^\infty \exp\left\{\kappa u - \tfrac{1}{2}(\kappa+u)^2 - au\right\} du$$

$$= \frac{1}{\kappa} \int_\eta^\infty \exp\left\{-\tfrac{1}{2}\kappa^2 - \tfrac{1}{2}u^2 - au\right\} du$$

$$\leqslant \frac{e^{-\frac{1}{2}\kappa^2}}{\kappa} \int_\eta^\infty e^{-\frac{1}{2}u^2} du$$

$$\leqslant \frac{e^{-\frac{1}{2}\kappa^2-\frac{1}{2}\eta^2}}{\kappa} \int_0^\infty e^{-\frac{1}{2}v^2} dv$$

$$= \frac{e^{-\frac{1}{2}\kappa^2-\frac{1}{2}\eta^2}}{\kappa} \left(\frac{\pi}{2}\right)^{\frac{1}{2}} = e^{-\frac{1}{2}\nu-\frac{1}{2}\eta^2}\left(\frac{\pi}{2\nu}\right)^{\frac{1}{2}},$$

which is very small compared with the contribution of the interval $-\eta \leqslant u \leqslant \eta$, where $\eta = \kappa^\epsilon$ $(0 < \epsilon < \tfrac{1}{3})$.

Lastly, the contribution to I of the interval $-\kappa \leqslant u \leqslant -\eta$, which is also positive, is equal to

$$\frac{1}{\kappa} \int_\eta^\kappa \exp\left\{\kappa^2 \log\left(1-\frac{v}{\kappa}\right) - \tfrac{1}{2}(\kappa-v)^2 + av\right\} dv$$

$$\leqslant \frac{1}{\kappa} \int_\eta^\kappa \exp\left\{-\tfrac{1}{2}\kappa^2 - \tfrac{1}{2}v^2 + av\right\} dv$$

$$= \frac{1}{\kappa} \exp\left\{\tfrac{1}{2}a^2 - \tfrac{1}{2}\kappa^2\right\} \int_\eta^\kappa \exp\left\{-\tfrac{1}{2}(v-a)^2\right\} dv$$

$$= \frac{1}{\kappa} \exp\left\{\tfrac{1}{2}a^2 - \tfrac{1}{2}\kappa^2\right\} \int_{\frac{1}{2}(\eta-a)^2}^{\frac{1}{2}(\kappa-a)^2} e^{-t} \frac{dt}{\sqrt{(2t)}}$$

$$\leqslant \frac{1}{\kappa} \exp\left\{\tfrac{1}{2}a^2 - \tfrac{1}{2}\kappa^2 - \tfrac{1}{2}(\eta-a)^2\right\} \int_0^\infty e^{-t} \frac{dt}{\sqrt{(2t)}}$$

$$= \left(\frac{\pi}{2\nu}\right)^{\frac{1}{2}} \exp\left\{\tfrac{1}{2}a^2 - \tfrac{1}{2}\nu - \tfrac{1}{2}(\eta-a)^2\right\},$$

which again is small compared with the contribution of the interval $-\eta \leqslant u \leqslant \eta$.

We have thus shown that, as $\nu \to +\infty$,

$$D_{-\nu-1}(a) \sim \frac{\sqrt{\pi}}{\Gamma(\nu+1)} e^{-\frac{1}{2}\nu - a\sqrt{\nu}} \nu^{\frac{1}{2}\nu},$$

or, using Stirling's formula,

$$D_{-\nu-1}(a) \sim \frac{e^{\frac{1}{2}\nu - a\sqrt{\nu}}}{2^{\frac{1}{2}} \nu^{\frac{1}{2}\nu + \frac{1}{2}}}.$$

Cherry stated that

$$D_{-\nu-1}(a) \sim \frac{e^{\frac{1}{2}\nu - a\sqrt{\nu}}}{2^{\frac{1}{2}} \nu^{\frac{1}{2}\nu + \frac{1}{2}}} \left\{ 1 + O\left(\frac{1}{\sqrt{\nu}}\right) \right\}$$

for complex ν and a, provided that $|\text{ph}\,\nu| \leqslant \frac{1}{2}\pi$.

CHAPTER 6

WATSON'S LEMMA

21. Laplace integrals

An integral of the form

$$f(z) = \int_0^\infty e^{-zt} \phi(t)\, dt \qquad (21.1)$$

is called a Laplace integral. We assume that $\phi(t)$ is integrable over any finite interval $0 \leqslant t \leqslant T$ and that the integral (21.1) is the limit as $T \to +\infty$ of the corresponding integral over this finite interval. It can be shown that, if the integral (21.1) converges for $z = z_0$, it is convergent in the half-plane $\mathscr{R}z > \mathscr{R}z_0$, and that it is uniformly convergent on any compact set in this half-plane. If α is the greatest lower bound of all such numbers $\mathscr{R}z_0$, the integral (21.1) is convergent for $\mathscr{R}z > \alpha$, divergent for $\mathscr{R}z < \alpha$. The constant α is called the abscissa of convergence, the half-plane $\mathscr{R}z > \alpha$ the half-plane of convergence. Evidently $f(z)$ is an analytic function, regular in the half-plane of convergence. If $\alpha = -\infty$, $f(z)$ is an integral function.

Since many of the special functions can be represented as Laplace integrals, it is desirable to be able to give an asymptotic representation of such an integral as $|z| \to \infty$ in the half-plane of convergence. If $\phi(t)$ has continuous derivatives of all orders, the process of integration by parts used in chapter 3 gives

$$f(z) \sim \sum_0^\infty \frac{\phi^{(n)}(0)}{z^{n+1}},$$

the result one would get if one substituted for $\phi(t)$ its Taylor expansion and integrated formally term by term. The justification of this formal process is provided by a lemma due to Watson [30]. This lemma also gives the asymptotic expansion in the case when the occurrence of a branch-point of $\phi(t)$ at the origin would cause the integration-by-parts method to fail.

22. Watson's lemma

Let $\phi(t)$ be an analytic function of t, regular, apart from a branch-point at 0, when $|t| \leqslant R+\delta$, $|\operatorname{ph} t| \leqslant \Delta < \pi$, where R, δ, Δ are positive; and let

$$\phi(t) = \sum_{m=1}^{\infty} a_m t^{(m/r)-1},$$

when $|t| \leqslant R$, r being positive. Also let $|\phi(t)| < K e^{bt}$, where K and b are positive numbers independent of t, when t is positive and $t \geqslant R$. Then

$$\int_0^{\infty} e^{-zt} \phi(t)\,dt \sim \sum_{m=1}^{\infty} a_m \Gamma\left(\frac{m}{r}\right) z^{-m/r}$$

as $|z| \to \infty$ in the sector $|\operatorname{ph} z| \leqslant \tfrac{1}{2}\pi - \epsilon < \tfrac{1}{2}\pi$.

If $r = 1$, $\phi(t)$ does not have a branch-point at the origin and the condition $|\operatorname{ph} t| \leqslant \Delta$ is not needed.

To prove the lemma, we observe that, having fixed a positive integer M, we can choose a constant C so that

$$\left| \phi(t) - \sum_{m=1}^{M-1} a_m t^{(m/r)-1} \right| \leqslant C\, e^{bt} t^{(M/r)-1},$$

when $t \geqslant 0$. Hence

$$\int_0^{\infty} e^{-zt} \phi(t)\,dt = \sum_{m=1}^{M-1} a_m \int_0^{\infty} e^{-zt} t^{(m/r)-1}\,dt + R_M$$

$$= \sum_{m=1}^{M-1} a_m \Gamma\left(\frac{m}{r}\right) z^{-m/r} + R_M,$$

where $\qquad R_M = \int_0^{\infty} e^{-zt}\left\{ \phi(t) - \sum_{m=1}^{M-1} a_m t^{(m/r)-1} \right\} dt.$

We have to show that $z^{M/r} R_M$ is bounded as $|z| \to \infty$ in the angle $|\operatorname{ph} z| \leqslant \tfrac{1}{2}\pi - \epsilon < \tfrac{1}{2}\pi$.

If we write $\mathscr{R}z = x$, we have

$$|R_M| \leqslant \int_0^{\infty} e^{-zt} C\, e^{bt} t^{(M/r)-1}\,dt$$

$$= \frac{C}{(x-b)^{M/r}} \Gamma\left(\frac{M}{r}\right),$$

when $x > b$. But since $|\mathrm{ph}\,z| \leqslant \frac{1}{2}\pi - \epsilon$, $x \geqslant |z| \sin \epsilon$ and so $x > b$ when $|z| > b\,\mathrm{cosec}\,\epsilon$. Hence, if $|\mathrm{ph}\,z| \leqslant \frac{1}{2}\pi - \epsilon$ and $|z| > b\,\mathrm{cosec}\,\epsilon$,

$$|z^{M/r} R_M| \leqslant \frac{C\,|z|^{M/r}}{\{|z|\sin\epsilon - b\}^{M/r}}\,\Gamma\!\left(\frac{M}{r}\right),$$

and this is bounded as $|z| \to \infty$.

The following alternative formulation of the lemma is sometimes useful.

Let $\phi(t)$ be an analytic function of t, regular in a neighbourhood of the origin, and let

$$\phi(t) = \sum_{0}^{\infty} a_m t^m$$

be its Taylor series. Also let

$$|\phi(t)| \leqslant K e^{bt^r},$$

where K and b are positive numbers independent of t, when $t \geqslant 0$, r being positive. Then

$$\int_{0}^{\infty} e^{-zt^r} \phi(t)\,dt \sim \sum_{m=0}^{\infty} \frac{a_m}{r}\,\Gamma\!\left(\frac{m+1}{r}\right) z^{-(m+1)/r}$$

as $|z| \to \infty$ in the sector $|\mathrm{ph}\,z| \leqslant \frac{1}{2}\pi - \epsilon < \frac{1}{2}\pi$.

23. The function $\Psi(1, 1; z)$

In section 7 we found the asymptotic expansion of the confluent hypergeometric function $\Psi(1,1;z)$ by the method of integration by parts, the expansion being valid as $|z| \to \infty$ in a sector $|\mathrm{ph}\,z| \leqslant \pi - \delta < \pi$. This function, which we denoted for brevity by $F(z)$, is defined by

$$F(z) = \int_{0}^{\infty} e^{-t}\,\frac{dt}{t+z}$$

and is regular in the complex plane cut along the negative real axis.

When z is positive, the substitution $z = tu$ gives

$$F(z) = \int_{0}^{\infty} e^{-zu}\,\frac{du}{1+u}, \tag{23.1}$$

and, by analytical continuation, this formula holds in the half plane $\mathscr{R}z > 0$. An application of Watson's lemma gives

$$F(z) \sim \sum_{1}^{\infty} \frac{(-1)^{n-1}(n-1)!}{z^n}, \qquad (23.2)$$

as $|z| \to \infty$ in the sector $|\mathrm{ph}\,z| \leqslant \frac{1}{2}\pi - \delta < \frac{1}{2}\pi$. This is the expansion (7.2), but proved here under more restricted conditions.

If we rotate the path of integration in (23.1) through an angle $\frac{1}{2}\pi$, we obtain another integral representation of $F(z)$, namely

$$F(z) = \int_{0}^{\infty} e^{-izv} \frac{i\,dv}{1+iv}$$

valid in the half plane $\mathscr{I}z < 0$. This again leads to the expansion (23.2) as $|z| \to \infty$, but valid now in the sector

$$-\pi + \delta \leqslant \mathrm{ph}\,z \leqslant -\delta < 0.$$

Similarly, a rotation through an angle $-\frac{1}{2}\pi$ leads to the same expression in $0 < \delta \leqslant \mathrm{ph}\,z \leqslant \pi - \delta$.

This device of rotating the path of integration often enables us to extend the range of values of $\mathrm{ph}\,z$ in which the asymptotic expansion of a Laplace integral holds. In the present example, the method of integration by parts is preferable.

24. The logarithm of the Gamma function

The asymptotic expansion of the logarithmic derivative of $\Gamma(p)$ can be obtained by a straightforward application of Watson's lemma.

From Euler's formula,

$$\Gamma(p) = \lim_{n \to \infty} \frac{n!}{p(p+1)(p+2)\ldots(p+n-1)} n^p,$$

we have, in the usual notation,

$$\psi(p) = \frac{\Gamma'(p)}{\Gamma(p)} = \lim_{n \to \infty} \left\{ \log n - \frac{1}{p} - \frac{1}{p+1} - \ldots - \frac{1}{p+n-1} \right\}$$

$$= \lim_{n \to \infty} \left\{ \int_{0}^{\infty} \frac{e^{-t} - e^{-nt}}{t}\,dt - \sum_{0}^{n-1} \int_{0}^{\infty} e^{-(p+r)t}\,dt \right\},$$

where $\mathscr{R}p > 0$. Hence

$$
\begin{aligned}
\psi(p) &= \lim_{n\to\infty}\left\{\int_0^\infty \frac{e^{-t}-e^{-nt}}{t}\,dt - \int_0^\infty \frac{1-e^{-nt}}{1-e^{-t}}\,e^{-pt}\,dt\right\}\\
&= \lim_{n\to\infty}\left\{\int_0^\infty \left(\frac{e^{-t}}{t}-\frac{e^{-pt}}{1-e^{-t}}\right)dt - \int_0^\infty \left(\frac{1}{t}-\frac{e^{-pt}}{1-e^{-t}}\right)e^{-nt}\,dt\right\}\\
&= \int_0^\infty \left(\frac{e^{-t}}{t}-\frac{e^{-pt}}{1-e^{-t}}\right)dt\\
&= \int_0^\infty \frac{e^{-t}-e^{-pt}}{t}\,dt + \int_0^\infty \left\{\frac{1}{t}-\frac{1}{1-e^{-t}}\right\}e^{-pt}\,dt\\
&= \log p + \int_0^\infty \left\{\frac{1}{t}-\frac{1}{1-e^{-t}}\right\}e^{-pt}\,dt,
\end{aligned}
$$

provided that $\mathscr{R}p > 0$.

Now
$$
\frac{1}{t}-\frac{1}{1-e^{-t}}
$$

is an analytic function of t, regular in $|t| < 2\pi$. In this region, it can be expanded as a Taylor series

$$
\sum_0^\infty \frac{B_{n+1}}{(n+1)!}\,(-1)^n t^n,
$$

where the coefficients B_n are the Bernoulli numbers defined by $B_n = B_n(0)$. Here $B_n(x)$ is Bernoulli's polynomial in the notation of Erdélyi, Magnus, Oberhettinger and Tricomi's *Higher Transcendental Functions*, which differs from that of Whittaker and Watson's *Modern Analysis*. The values of B_n are connected by a recurrence relation; the first few are

$$
B_1 = -\tfrac{1}{2},\quad B_2 = \tfrac{1}{6},\quad B_4 = -\tfrac{1}{30},\quad B_6 = \tfrac{1}{42},
$$

and $B_{2n+1} = 0$ for $n > 0$. In this notation,

$$
\frac{1}{t}-\frac{1}{1-e^{-t}} = -\tfrac{1}{2}-\sum_1^\infty \frac{B_{2m}}{(2m)!}\,t^{2m-1}.
$$

Since the expression on the left-hand side of this equation is bounded when t is positive, Watson's lemma is applicable and gives

$$
\psi(p) \sim \log p - \frac{1}{2p} - \sum_0^\infty \frac{B_{2m}}{2m}\frac{1}{p^{2m}}
$$

as $|p| \to \infty$ in $|\mathrm{ph}\,p| \leqslant \tfrac{1}{2}\pi - \delta < \tfrac{1}{2}\pi$.

It follows by integration that

$$\log \Gamma(p) \sim (p - \tfrac{1}{2}) \log p - p + \sum_{1}^{\infty} \frac{B_{2m}}{2m(2m-1)} \frac{1}{p^{2m}} + C,$$

where C is a constant. That $C = \tfrac{1}{2} \log (2\pi)$ is proved most simply by substituting the asymptotic series into Legendre's duplication formula

$$\log \Gamma(p) + \log \Gamma(p + \tfrac{1}{2}) + (2p-1) \log 2 = \log \Gamma(2p) + \tfrac{1}{2} \log \pi.$$

Although the formulae have been proved here only in a sector inside the right-hand half-plane, they do, in fact, hold in $|\mathrm{ph}\, p| \leqslant \pi - \delta < \pi$. For, by a rotation of the path of integration through a positive or negative acute angle α, we can show that the formulae hold also when $|\mathrm{ph}\, (p\, e^{\alpha i})| \leqslant \tfrac{1}{2}\pi - \delta < \tfrac{1}{2}\pi$, from which the result follows.

25. The Gamma function, continued

If p is real and positive,

$$\Gamma(p) = \frac{1}{p} \Gamma(p+1) = \frac{1}{p} \int_0^{\infty} e^{-x} x^p \, dx = p^p \int_0^{\infty} e^{-pw} w^p \, dw.$$

By analytical continuation, this formula holds when $\mathscr{R}p > 0$, it being understood that p^p means $e^{p \log p}$, where $\log p$ has its principal value. Thus we have

$$p^{-p} \Gamma(p) = \int_0^{\infty} (w\, e^{-w})^p \, dw.$$

From this formula we deduce the asymptotic expansion of $\Gamma(p)$ as $|p| \to \infty$ in the sector $|\mathrm{ph}\, p| \leqslant \tfrac{1}{2}\pi - \epsilon < \tfrac{1}{2}\pi$, and then show that the expansion is in fact valid when $|\mathrm{ph}\, p| \leqslant \pi - \epsilon < \pi$.

The function $w\, e^{-w}$ increases steadily from 0 to e^{-1} as w increases from 0 to 1, and then decreases steadily from e^{-1} to 0 as w increases from 1 to $+\infty$. We therefore write

$$p^{-p} \Gamma(p) = \int_0^1 (w_2\, e^{-w_2})^p \, dw_2 + \int_1^{\infty} (w_1\, e^{-w_1})^p \, dw_1,$$

it being convenient to use different symbols for w in the two ranges of integration. If we put $w_1 e^{-w_1} = e^{-1-\tau}$, τ increases steadily from 0 to $+\infty$ as w_1 increases from 1 to $+\infty$; and if we put $w_2 e^{-w_2} = e^{-1-\tau}$, τ decreases steadily from $+\infty$ to 0 as w_2 increases from 0 to 1. Changing the variable of integration in this way, we obtain

$$p^{-p} e^p \, \Gamma(p) = \int_0^\infty e^{-p\tau} \left(\frac{dw_1}{d\tau} - \frac{dw_2}{d\tau} \right) d\tau,$$

to which we apply Watson's lemma. The proof is based on a method used by Watson [31] in solving a related problem of Ramanujan.

When $\tau \geqslant 0$, the functions w_1 and w_2 are the two real solutions of the equation

$$w - \log w = 1 + \tau,$$

which take the value 1 when $\tau = 0$. If we put $w = 1 + W$, $\tau = \frac{1}{2}z^2$, the equation becomes $F(W, z) = 0$ where

$$F(W, z) = W - \log(1+W) - \tfrac{1}{2}z^2;$$

and we now regard W and z as complex variables.

Now in order that the equation $F = 0$ may have a unique solution W which takes the value W_0 at $z = z_0$ and is regular in a neighbourhood of z_0, it is necessary and sufficient that $\partial F/\partial W$ does not vanish when $z = z_0$, $W = W_0$. But

$$\frac{\partial F}{\partial W} = \frac{W}{1+W},$$

which vanishes only if $W = 0$, the corresponding values of z being given by $z^2 = 4n\pi i$, where n is an integer or zero. The equation $F = 0$ thus defines a many-valued function $W(z)$ with branch-points at $z = 0$ and at $z = \pm 2 \sqrt{(n\pi)} e^{\pm\frac{1}{4}\pi i}$, where the integer n is positive.

Near $W = 0$,

$$F = \tfrac{1}{2}W^2 - \tfrac{1}{3}W^3 + \tfrac{1}{4}W^4 - \ldots - \tfrac{1}{2}z^2,$$

so that z, regarded as a function of W, has two branches

$$z = \pm \, W(1 - \tfrac{2}{3}W + \tfrac{2}{4}W^2 - \ldots)^{\frac{1}{2}}.$$

Taking the upper sign, we have

$$z = W(1 - \tfrac{1}{3}W + \ldots),$$

where the expression on the right-hand side is an analytic function of W which is regular when $|W| < 1$ and has a simple zero at $W = 0$. This last equation has, therefore, a unique solution $W_1(z)$ which vanishes when $z = 0$ and is regular in a neighbourhood of $z = 0$. This solution is of the form

$$W_1 = \sum_1^\infty a_n z^n,$$

where $a_1 = 1$, $a_2 = \frac{1}{3}$. Similarly,

$$z = -W(1 - \tfrac{2}{3}W + \tfrac{2}{4}W^2 - \ldots)^{\frac{1}{2}}$$

has the unique solution

$$W_2 = \sum_1^\infty a_n(-z)^n,$$

which vanishes at $z = 0$ and is regular in a neighbourhood of the origin. We return later to the determination of the other coefficients.

Since

$$\frac{dW}{dz} = \frac{z(1+W)}{W},$$

the only singularities of W, regarded as a function of z, are the branch-points $\pm 2\sqrt{(n\pi)}\, e^{\pm\frac{1}{4}\pi i}$ already found, and so the series for W_1 and W_2 are convergent for $|z| < 2\sqrt{\pi}$.

If we put $z^2 = 2\tau$, we find that

$$w_1 = 1 + \sum_1^\infty a_n(2\tau)^{\frac{1}{2}n},$$

$$w_2 = 1 + \sum_1^\infty (-1)^n a_n(2\tau)^{\frac{1}{2}n},$$

the distinction between w_1 and w_2 being obtained from the fact that, when $\tau > 0$, w_1 is increasing but w_2 decreasing near $\tau = 0$. Apart from the branch-point at $\tau = 0$, w_1 and w_2 are regular functions of the *complex* variable τ in $|\tau| < 2\pi$.

Since

$$\frac{dw}{d\tau} = \frac{w}{w-1}$$

we have $\dfrac{dw_1}{d\tau} - \dfrac{dw_2}{d\tau} = \dfrac{w_1}{w_1-1} - \dfrac{w_2}{w_2-1} = \dfrac{1}{w_1-1} + \dfrac{1}{1-w_2}$,

which is bounded in $\tau \geqslant \tau_0$, for any positive value of τ_0.

The conditions of Watson's lemma are thus satisfied, and we obtain the asymptotic expansion of

$$\int_0^\infty e^{-p\tau} \left(\frac{dw_1}{d\tau} - \frac{dw_2}{d\tau} \right) d\tau$$

by substituting the series for w_1 and w_2 and integrating term by term. It follows that

$$\Gamma(p) \sim e^{-p} p^p \sqrt{\left(\frac{2\pi}{p}\right)} \left\{ 1 + \frac{3a_3}{p} + \frac{3.5a_5}{p^2} + \frac{3.5.7a_7}{p^3} + \ldots \right\}$$

where $|p|$ is large and $|\mathrm{ph}\, p| \leqslant \frac{1}{2}\pi - \epsilon < \frac{1}{2}\pi$.

In order to determine the coefficients a_n, we substitute the series $\sum\limits_1^\infty a_n z^n$ for W_1 in the equation

$$W_1 \frac{dW_1}{dz} = z(1 + W_1),$$

and equate coefficients, a process which is valid since all the series which occur are absolutely convergent in $|z| < 2\sqrt{\pi}$. From the identity

$$\sum_1^\infty a_m z^m \sum_1^\infty a_n n z^{n-1} = z + \sum_1^\infty a_n z^{n+1},$$

we thus obtain

$$a_{n-1} = \sum_{r+s=n+1} r a_r a_s = \sum_{r+s=n+1} s a_s a_r,$$

and so

$$2a_{n-1} = (n+1) \sum_{r+s=n+1} a_r a_s,$$

when $n \geqslant 2$. We know that $a_1 = 1$. When $n = 2$, we have $2a_1 = 6a_1 a_2$ so that $a_2 = \frac{1}{3}$. When $n \geqslant 3$,

$$2a_{n-1} = 2(n+1) a_1 a_n + (n+1) \sum_{r=2}^{n-1} a_r a_{n+1-r},$$

whence

$$a_n = \frac{a_{n-1}}{n+1} - \frac{1}{2} \sum_{r=2}^{n-1} a_r a_{n+1-r}.$$

This enables us to find the successive coefficients, of which the first few are

$$a_1 = 1, \quad a_2 = \tfrac{1}{3}, \quad a_3 = \tfrac{1}{36}, \quad a_4 = -\tfrac{1}{270}, \quad a_5 = \tfrac{1}{4320}.$$

There is no simple formula for the general coefficient, but Watson
has shown that, when n is large,

$$a_n \sim (-1)^n \frac{\cos \frac{1}{4}(n+1)\pi}{2^{n-\frac{3}{2}}n^{\frac{3}{2}}\pi^{\frac{1}{2}n}}.$$

Inserting the values of the coefficients, we have finally

$$\Gamma(p) \sim e^{-p}p^p \sqrt{\left(\frac{2\pi}{p}\right)} \left\{1 + \frac{1}{12p} + \frac{1}{288p^2} + \cdots\right\},$$

when $|p|$ is large and $|\mathrm{ph}\,p| \leqslant \frac{1}{2}\pi - \epsilon < \frac{1}{2}\pi$.

This asymptotic expansion actually holds when

$$|\mathrm{ph}\,p| \leqslant \pi - \epsilon < \pi.$$

A detailed proof could be rather tedious, but the esse ntial ide
is simple. We start with the formula

$$p^{-p}\,e^p\,\Gamma(p) = \int_0^\infty e^{-p\tau}F(\tau)\,d\tau,$$

where
$$F(\tau) = \frac{dw_1}{d\tau} - \frac{dw_2}{d\tau} = \frac{1}{w_1 - 1} - \frac{1}{w_2 - 1}.$$

It can be shown, when τ is complex, that $w_1 = O(|\tau|)$, $w_2 = O(|e^{-\tau}|)$
when $\mathscr{R}\tau > 0$. Hence $F(\tau)$ is bounded as $|\tau| \to \infty$, uniformly with
respect to $\mathrm{ph}\,\tau$ in any angle $-\frac{1}{2}\pi < \alpha \leqslant \mathrm{ph}\,\tau \leqslant \beta < \frac{1}{2}\pi$. As the
only singularities of $F(\tau)$ are branch points at the origin and
at points on the imaginary axis, we can rotate the path of integration through any positive or negative acute angle γ. Then

$$\int_0^\infty e^{-p\tau}F(\tau)\,d\tau = \int_0^\infty e^{-q\tau}F(\tau\,e^{i\gamma})\,e^{i\gamma}\,d\tau$$

provided that the real parts of p and of $q = p\,e^{i\gamma}$ are positive.
Hence by analytical continuation,

$$p^{-p}\,e^p\,\Gamma(p) = \int_0^\infty e^{-q\tau}F(\tau\,e^{i\gamma})\,e^{i\gamma}\,d\tau,$$

when $\mathscr{R}q > 0$. A repetition of the previous argument shows that

$$p^{-p}\,e^p\,\Gamma(p) \sim \left(\frac{2\pi\,e^{i\gamma}}{q}\right)^{\frac{1}{2}} \left\{1 + \frac{e^{i\gamma}}{12q} + \frac{e^{2i\gamma}}{288q^2} + \cdots\right\}$$

$$= \left(\frac{2\pi}{p}\right)^{\frac{1}{2}} \left\{1 + \frac{1}{12p} + \frac{1}{288p^2} + \cdots\right\},$$

E

the same series as before. But it has now been shown to be valid
as $|q| \to \infty$ in $|\mathrm{ph}\, q| \leqslant \frac{1}{2}\pi - \epsilon < \frac{1}{2}\pi$, that is, when the phase of p lies
between $-\frac{1}{2}\pi + \epsilon - \gamma$ and $\frac{1}{2}\pi - \epsilon - \gamma$. Since γ is any positive or
negative acute angle, the expansion is valid in $|\mathrm{ph}\, p| \leqslant \pi - \epsilon < \pi$.

26. The logarithmic integral

When x is positive, the logarithmic integral $\mathrm{li}\,(x)$ is defined by

$$\mathrm{li}\,(x) = \int_0^x \frac{dt}{\log t},$$

the integral being a Cauchy principal value when $x > 1$. If we
put $x = e^a$, $t = e^{a-v}$, this becomes

$$\mathrm{li}\,(e^a) = e^a P \int_0^\infty e^{-v} \frac{dv}{a-v},$$

when $a > 0$. This is the integral discussed in section 9. We now
apply Laplace's method in order to prove Stieltjes's result con-
cerning the best approximation by an asymptotic expansion to
which we referred earlier.

In the equation

$$e^{-a}\,\mathrm{li}\,(e^a) = P \int_0^\infty e^{-at} \frac{dt}{1-t},$$

put

$$\frac{1}{1-t} = 1 + t + t^2 + \ldots + t^{n-1} + \frac{t^n}{1-t}.$$

Then

$$e^{-a}\,\mathrm{li}\,(e^a) = \sum_{k=1}^n \int_0^\infty e^{-at} t^{k-1}\, dt + P \int_0^\infty e^{-at} \frac{t^n}{1-t}\, dt$$

$$= \sum_{k=1}^n \frac{(k-1)!}{a^k} + R_n(a),$$

where

$$R_n(a) = P \int_0^\infty e^{-at} \frac{t^n}{1-t}\, dt.$$

It can be shown that $R_n(a)$ is $O(1/a^{n+1})$ as $a \to \infty$, and so

$$e^{-a}\,\mathrm{li}\,(e^a) \sim \sum_{k=1}^\infty \frac{(k-1)!}{a^k},$$

the result obtained earlier with a different form for the remainder.
If $a = N + \eta$, where N is a large integer and $0 \leqslant \eta < 1$, the

smallest term in the asymptotic expansion is the Nth term. Stieltjes showed that the remainder $R_n(N+\eta)$ changes sign near $n = N$, so that the partial sum up to about the Nth term gives the best approximation.

By the definition of the Cauchy principal value,

$$R_N(N+\eta) = \lim_{\epsilon \to 0} \left(\int_0^{1-\epsilon} + \int_{1+\epsilon}^\infty e^{-(N+\eta)t} \frac{t^N}{1-t} dt \right)$$

$$= \lim_{\epsilon \to 0} (I_1 + I_2).$$

Introducing the notation of the preceding section, we have

$$I_1 = \int_0^{1-\epsilon} \frac{e^{-\eta w_2}}{1-w_2} (w_2 e^{-w_2})^N dw_2$$

$$= -\int_{\frac{1}{2}\epsilon^2}^\infty \frac{e^{-\eta w_2}}{1-w_2} (w_2 e^{-w_2})^N \frac{dw_2}{d\tau} d\tau,$$

$$I_2 = \int_{1+\epsilon}^\infty \frac{e^{-\eta w_1}}{1-w_1} (w_1 e^{-w_1})^N dw_1$$

$$= \int_{\frac{1}{2}\epsilon^2}^\infty \frac{e^{-\eta w_1}}{1-w_1} (w_1 e^{-w_1})^N \frac{dw_1}{d\tau} d\tau.$$

Actually the lower limits should not be $\frac{1}{2}\epsilon^2$; they should be different and of the form $\frac{1}{2}\epsilon^2 + O(\epsilon^3)$, but this does not affect the proof—it merely involves a slight irrelevant change in the definition of the Cauchy principal value. We now make $\epsilon \to 0$, and obtain

$$R_N(N+\eta) = \int_0^\infty e^{-N(1+\tau)} \left\{ \frac{e^{-\eta w_1}}{1-w_1} \frac{dw_1}{d\tau} - \frac{e^{-\eta w_2}}{1-w_2} \frac{dw_2}{d\tau} \right\} d\tau.$$

The conditions of Watson's lemma are satisfied; it remains to calculate the coefficients in the expansion near the origin of the expression in brackets under the sign of integration and then integrate term by term.

Taking only the dominant term, we have

$$R_N(N+\eta) = \sqrt{2} \, e^{-N-\eta} (\eta - \tfrac{1}{3}) \int_0^\infty e^{-N\tau} \tau^{-\frac{1}{2}} d\tau \left[1 + O\left(\frac{1}{N}\right) \right]$$

$$= e^{-N-\eta} \left(\frac{2\pi}{N}\right)^{\frac{1}{2}} (\eta - \tfrac{1}{3}) \left[1 + O\left(\frac{1}{N}\right) \right],$$

which is equivalent to

$$R_N(N+\eta) = \frac{N!}{(N+\eta)^{N+1}}\left\{\eta - \tfrac{1}{3} + O\!\left(\frac{1}{N}\right)\right\},$$

as $N \to \infty$. Stieltjes, whose proof was somewhat different, gave very complicated formulae for the coefficients of $1/N$ and $1/N^2$.

It follows that

$$R_{N-1}(N+\eta) = \frac{N!}{(N+\eta)^{N+1}}\left\{\eta + \tfrac{2}{3} + O\!\left(\frac{1}{N}\right)\right\},$$

$$R_{N+1}(N+\eta) = \frac{N!}{(N+\eta)^{N+1}}\left\{\eta - \tfrac{4}{3} + O\!\left(\frac{1}{N}\right)\right\},$$

and so $R_n(N+\eta)$ changes sign between $n = N-1$ and $n = N+1$ if $0 \leqslant \eta < 1$.

27. An application of Watson's lemma to a loop integral

Tricomi and Erdélyi [28] have pointed out that Watson's lemma can be applied to loop integrals of the Laplace type, and used this idea to find the asymptotic expansion of

$$\Gamma(\nu+\alpha)/\Gamma(\nu+\beta)$$

when $|\nu|$ is large.

They started with Euler's integral

$$\frac{\Gamma(\nu+a)\,\Gamma(\beta-\alpha)}{\Gamma(\nu+\beta)} = \int_0^1 t^{\nu+\alpha-1}(1-t)^{\beta-\alpha-1}\,dt, \qquad (27.1)$$

the many-valued functions having their principal values. This result is true under two conditions. The first is that $\mathscr{R}(\nu+\alpha) > 0$; if $|\mathrm{ph}\,\nu| \leqslant \tfrac{1}{2}\pi - \delta < \tfrac{1}{2}\pi$, we can choose $|\nu|$ so large that this condition is satisfied. The second is that $\mathscr{R}(\beta-\alpha) > 0$; it was to get rid of this restriction that Tricomi and Erdélyi introduced a loop integral.

If we put $t = e^{-r}$ in (27.1) we obtain a Laplace integral

$$\frac{\Gamma(\nu+\alpha)\,\Gamma(\beta-\alpha)}{\Gamma(\nu+\beta)} = \int_0^\infty e^{-\nu r}e^{-\alpha r}(1-e^{-r})^{\beta-\alpha-1}\,dr,$$

from which it follows that

$$\frac{\Gamma(\nu+\alpha)\,\Gamma(\beta-\alpha)}{\Gamma(\nu+\beta)} = \frac{1}{2i\sin\pi(\beta-\alpha)}\int_{-\infty}^{(0+)} e^{\nu z}\phi(z)\,dz, \qquad (27.2)$$

where $$\phi(z) = e^{\alpha z}(e^z - 1)^{\beta-\alpha-1}.$$

In (27.2), $|\mathrm{ph}\,z| \leqslant \pi$ and $\phi(z)$ has its principal value. It is assumed that $\beta - \alpha$ is not an integer.

Watson's lemma can be applied directly to (27.2). For it is clear that the proof still applies, provided that the restriction on the growth of $\phi(z)$ is applied along the whole loop, and that a power series expansion of $\phi(z)$ is valid on the loop in a neighbourhood of the origin. Both assumptions hold here. The expansion of $\phi(z)$, valid when $|z| < 2\pi$, is

$$\phi(z) = \sum_0^\infty B_n^{(\alpha-\beta+1)}(\alpha) \frac{z^{\beta-\alpha+n-1}}{n!}, \qquad (27.3)$$

where $B_n^{(\kappa)}(\alpha)$ is Nörlund's generalized Bernoulli polynomial.

Now, by Hankel's loop integral for the Gamma function

$$\Gamma(\beta-\alpha+n) = \frac{1}{2i \sin \pi(\beta-\alpha+n)} \int_{-\infty}^{(0+)} e^t t^{\beta-\alpha+n-1} dt$$

$$= \frac{\nu^{\beta-\alpha+n}}{2i \sin \pi(\beta-\alpha)} (-1)^n \int_{-\infty}^{(0+)} e^{\nu z} z^{\beta-\alpha+n-1} dz,$$

when ν is positive. By analytical continuation, this formula also holds when $\mathscr{R}\nu$ is positive. Hence, if we substitute the series (27.3) in (27.2) and integrate term by term, we obtain

$$\frac{\Gamma(\nu+\alpha)}{\Gamma(\nu+\beta)} \sim \sum_0^\infty \frac{(-1)^n}{n!} B_n^{(\alpha-\beta+1)}(\alpha) \frac{\Gamma(\beta-\alpha+n)}{\Gamma(\beta-\alpha)} \frac{1}{\nu^{\beta-\alpha+n}}, \quad (27.4)$$

as $|\nu| \to \infty$ in $|\mathrm{ph}\,\nu| \leqslant \frac{1}{2}\pi - \delta < \frac{1}{2}\pi$.

It can be shown that (27.2) is valid when ν is positive if the lower limit is replaced by $-\infty e^{i\gamma}$, where γ is a positive or negative acute angle. With this change, (27.2) holds when

$$|\mathrm{ph}\,(\nu e^{i\gamma})| \leqslant \frac{1}{2}\pi - \delta.$$

A repetition of the proof then shows that (27.4) is also valid in this new angle; and, by varying γ, it follows that (27.4) holds when $|\mathrm{ph}\,\nu| \leqslant \pi - \delta < \pi$.

This asymptotic formula is true despite the fact that

$$\Gamma(\nu+\alpha)/\Gamma(\nu+\beta)$$

has poles at $\nu = -\alpha, -\alpha-1, -\alpha-2, \ldots$; if $|\nu|$ is sufficiently large, none of these poles lie in the fixed angle $|\mathrm{ph}\,\nu| \leqslant \pi - \delta$. But the approximation will break down if it is used for moderately large values of $|\nu|$ if ν is near one of these poles. This difficulty can be overcome by writing

$$\frac{\Gamma(\nu+\alpha)}{\Gamma(\nu+\beta)} = \frac{\sin \pi(\nu+\beta)}{\sin \pi(\nu+\alpha)} \frac{\Gamma(1-\beta-\nu)}{\Gamma(1-\alpha-\nu)},$$

and using the asymptotic expansion of the expression on the right-hand side for $|\mathrm{ph}(-\nu)| \leqslant \pi - \delta$.

Lastly, the restriction that $\beta - \alpha$ is not an integer is of little importance; for if $\beta - \alpha$ is an integer, $\Gamma(\nu+\alpha)/\Gamma(\nu+\beta)$ is a rational function.

The first two terms of (27.4) give

$$\frac{\Gamma(\nu+\alpha)}{\Gamma(\nu+\beta)} = \nu^{\alpha-\beta}\left[1 + \frac{(\alpha-\beta)(\alpha+\beta-1)}{2\nu} + O\left(\frac{1}{|\nu|^2}\right)\right].$$

An interesting particular case of (27.4) is

$$\frac{\Gamma(\nu+\tfrac{1}{2})}{\Gamma(\nu+1)} = \frac{1}{\sqrt{\nu}}\left[1 - \frac{1}{8\nu} + \frac{1}{128\nu^2} + O\left(\frac{1}{|\nu|^3}\right)\right],$$

which gives

$$\frac{1.3.5\ldots(2n-1)}{2.4.6\ldots(2n)} = \frac{1}{\sqrt{(\pi n)}}\left[1 - \frac{1}{8n} + \frac{1}{128n^2} + O\left(\frac{1}{n^3}\right)\right],$$

when n is a large positive integer.

CHAPTER 7

THE METHOD OF STEEPEST DESCENTS

28. The origin of the method

The method of steepest descents goes back to a posthumous paper of Riemann [25] in which he found an asymptotic approximation to the hypergeometric function

$$F(n-c, n+a+1; \ 2n+a+b+2; \ s)$$

when n is large and positive. This function is a multiple of

$$\int_0^1 z^{n+a}(1-z)^{n+b}(1-sz)^{c-n}\,dz,$$

and is an analytic function, regular in the complex s plane cut along the real axis from $s = 1$ to $s = +\infty$.

The integral may be written in the form

$$\int_0^1 \{f(z)\}^n \, \phi(z)\,dz, \tag{28.1}$$

where

$$f(z) = z(1-z)(1-sz)^{-1}, \quad \phi(z) = z^a(1-z)^b(1-sz)^c.$$

Riemann started by considering the curves in the complex z plane on which $|f(z)|$ is constant. These are the level curves of the surface whose equation is $t = |f(x+iy)|$, in a space in which (x, y, t) are rectangular Cartesian co-ordinates. If the modulus is small, the level curve has two branches, approximately non-intersecting circles of small radius with centres at $z = 0$ and $z = 1$. As the modulus increases, these branches approach one another and for one value of the modulus form a single curve with double point at a point $z = \zeta$. If the modulus is large, the level curve has again two branches, approximately a small circle with centre at $z = 1/s$ and a large circle with centre at the origin. As the modulus is decreased, the two branches approach each other and, for one value of the modulus, form a simple curve with double point at

$z = \zeta'$. By the Cauchy–Riemann equations, these double-points satisfy $f'(z) = 0$, and so are

$$\zeta = \frac{1}{1 + \sqrt{(1-s)}}, \quad \zeta' = \frac{1}{1 - \sqrt{(1-s)}},$$

the branch of the square root having positive real part in the cut s plane.

The points $z = \zeta$ and $z = \zeta'$ are saddle-points or cols on the surface $t = |f(x+iy)|$. Now $f(0) = f(1) = 0$; but since $f(\zeta) = \zeta^2$, $|f(\zeta)| > 0$, and so there are paths on this surface from 0 to ζ on which $|f(z)|$ increases steadily, and thence to 1 on which $|f(z)|$ decreases steadily. More picturesquely, there are uphill paths from 0 to ζ in the valley below the saddle-point and downhill paths from ζ to 1. It was such a path that Riemann took as his path of integration.

To get the desired asymptotic approximation, Riemann, using essentially Laplace's method, said that the dominant part of the integral, when n is large, was given by the neighbourhood of the saddle-point; and, near the saddle-point, he put

$$\{f(z)\}^n = \zeta^{2n} e^{-n\tau^2}, \tag{28.2}$$

where τ is real. This meant that he integrated, as we shall see, along the steepest direction at the saddle-point. But equation (28.2), when τ varies from $-\infty$ to $+\infty$, defines a curve, starting at $z = 0$ and ending at $z = 1$, which is the steepest path in the two valleys. If Riemann had used the whole path, he would have obtained a complete asymptotic expansion, instead of an asymptotic approximation.

It was Debye [10], in his work on the asymptotic expansions of the Bessel functions, who first realized that, by using the whole path and avoiding Laplace's method, one could obtain complete asymptotic expansions.

We distinguish between the two methods by calling Riemann's method of using only the neighbourhood of the saddle-point *the saddle-point method*, and Debye's method *the method of steepest descents*.

It is an advantage to make a slight change in notation. Instead of considering integrals of the form (28.1), we put $f(z) = e^{w(z)}$ and consider

$$\int e^{nw(z)} \phi(z)\, dz.$$

The curves on which $|f(z)|$ is constant are those on which $\mathscr{R}w(z)$ is constant. It is therefore more convenient to consider the surface $u = \mathscr{R}w(x+iy)$ instead of $t = |f(x+iy)|$, (x,y,u) being rectangular Cartesian co-ordinates.

29. Debye's method of steepest descents

The method of steepest descents consists essentially in choosing a path of integration with a particular geometrical property. Let us suppose that we wish to find the asymptotic expansion, when ν is large and positive, of a function defined by an integral of the form

$$\int e^{\nu w(z)} \phi(z)\, dz, \qquad (29.1)$$

where the path of integration is an arc or a closed curve in the z plane. The functions $w(z)$ and $\phi(z)$ are independent of ν and are analytic functions of z, regular in a domain which contains the path of integration. The idea is to deform the path of integration to satisfy the following conditions:

(i) the path passes through a zero z_0 of $w'(z)$;

(ii) the imaginary part of $w(z)$ is constant on the path.

If we write

$$z = x+iy, \quad w(z) = u(x,y) + iv(x,y), \qquad (29.2)$$

where x, y, u, v are real, the equation of the new path of integration is $v(x,y) = v(x_0, y_0)$, and the integrand becomes

$$e^{i\nu v(x_0,\, y_0)}\, e^{\nu u(x,\, y)}\, \phi(x+iy).$$

Since ν and u are real and ϕ does not depend on ν, the integrand does not oscillate rapidly on the new path for large values of ν.

To obtain a geometrical picture of the new path, consider the surface S whose equation is $u = u(x,y)$ in rectangular Cartesian co-ordinates, the axis of u being vertically upwards. By the Cauchy–Riemann conditions,

$$w'(z) = \frac{\partial u}{\partial x} - i \frac{\partial u}{\partial y}.$$

Hence, if z_0 is a zero of $w'(z)$, the tangent plane to S at the corresponding point (x_0, y_0, u_0) is horizontal. But, since

$$\frac{\partial^2 u}{\partial x^2} + \frac{\partial^2 u}{\partial y^2} = 0,$$

this point must be a saddle-point, not a maximum or minimum.

The shape of the surface S can be represented on the (x, y)-plane by drawing the contour lines or level curves on which u is constant; a saddle-point is evidently a multiple point on a particular level curve. The level curve through a saddle-point separates the nearby part of S into valleys below the saddle-point and rising ground above the saddle-point. The curves $v = $ constant are the orthogonal trajectories of the level curves, and so are the maps on the (x, y)-plane of the steepest curves on S. So a Debye curve is a steepest path through a saddle-point. In the simplest case, when the saddle-point is a double-point of the corresponding level curve, there are two paths of steepest descent from it and two paths of steepest ascent.

On a steepest path through a saddle-point z_0, we have

$$w(z) = w(z_0) - \tau, \tag{29.3}$$

where τ is real, and so, if s is the arc of the path, $d\tau/ds = \pm |w'(z)|$. Therefore $d\tau/ds$ can only change sign if the path goes through another saddle-point or a singularity of $w'(z)$, but it is rare for this to happen. The variable τ is usually monotonic on a steepest path from a saddle-point and either increases to $+\infty$ or decreases to $-\infty$. But since the integrand is

$$e^{\nu w(z_0) - \nu \tau} \phi,$$

a path on which $\tau \to -\infty$ would lead to a divergent integral. Hence we try to choose paths of integration on which τ is positive —these are the paths of steepest descent from the saddle-point. If it is possible to deform the path of integration and express the integral as the sum of integrals along paths of steepest descent from a saddle-point, all that remains is to consider the asymptotic behaviour of integrals of the form

$$e^{\nu w(z_0)} \int_0^\infty e^{-\nu \tau} \phi \frac{dz}{d\tau} \, d\tau, \tag{29.4}$$

where ν is large and positive, and, to each of these, we can usually apply Watson's lemma.

Generally, it is rather difficult to get from (29.3) a parametric equation for each of the paths of steepest descent from a saddle-point, though it is quite easy to find the form near the saddle-point. Let us consider the case when $w'(z)$ has a zero of order $m-1$ at a saddle-point z_0 so that

$$w(z) = w(z_0) - (z-z_0)^m f(z), \tag{29.5}$$

where $f(z)$ is regular in some neighbourhood of z_0 and $f(z_0) = a e^{-\alpha i}$ where $a > 0$. Near the saddle-point the level curves and steepest curves will be roughly the same as for

$$w(z) = w(z_0) - a e^{-\alpha i}(z-z_0)^m.$$

Hence, if $z = z_0 + r e^{\theta i}$,

$$u = u_0 - ar^m \cos(m\theta - \alpha), \quad v = v_0 - ar^m \sin(m\theta - \alpha).$$

The level curves $u = u_0$ are approximately $\theta = \{(n+\frac{1}{2})\pi + \alpha\}/m$ and the steepest curves $v = v_0$ are $\theta = (n\pi + \alpha)/m$, where $n = 1, 2, 3, \ldots, 2m$. There are thus $2m$ steepest directions from z_0, m directions of steepest ascent and m directions of steepest descent. The level curves near z_0 divide S into m valleys below the saddle-point and m hills above the saddle-point. The paths of integration are in the valleys. Figures 1 and 2 show the lie of the land near a saddle-point in the simplest cases $m = 2$, $m = 3$ with $\alpha = 0$; the valleys are shaded, the steepest curves dotted, and the unbroken lines the level curves.

By (29.3) and (29.5) we have to solve the equation

$$(z-z_0)^m f(z) = \tau, \tag{29.6}$$

where τ is positive on a path of steepest descent. It is convenient to regard τ as a complex variable and use the theory of analytic functions to find the m solutions of (29.6) which have the value z_0 when $\tau = 0$.

By hypothesis, $f(z)$ can be expanded as a convergent power series

$$f(z) = a_0 + a_1(z-z_0) + a_2(z-z_0)^2 + \ldots,$$

where $a_0 \neq 0$. By continuity, there is a neighbourhood of z_0 in which $f(z)$ does not vanish. The principal value $g(z)$ of the mth

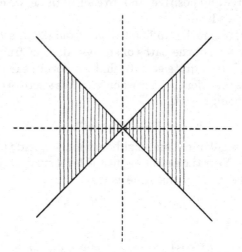

Figure 1. $m = 2, \alpha = 0$.

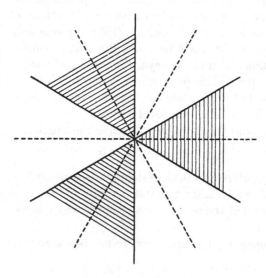

Figure 2. $m = 3, \alpha = 0$.

root of $f'(z)$ is therefore analytic in some neighbourhood of z_0 and can be expanded as a power series

$$g(z) = b_0 + b_1(z - z_0) + b_2(z - z_0)^2 + \ldots,$$

where b_0 is the principal value of $a_0^{1/m}$. The equation (29.6) then becomes

$$(z - z_0) g(z) = t,$$

where t is an mth root of τ. But by the inverse function theorem for analytic functions this equation has a unique solution

$$z = z_0 + c_1 t + c_2 t^2 + c_3 t^3 + \ldots, \qquad (29.7)$$

which takes the value z_0 when $t = 0$ and is regular in a neighbourhood of $t = 0$. This gives one solution of (29.6); the others are obtained by replacing t by ωt, where ω is a complex mth root of unity. When t is real, (29.7) gives the parametric equation of a steepest path through the saddle-point; when $t > 0$, it is a path of steepest descent.

In order to find a complete asymptotic expansion, it is necessary first to find the coefficients A_n in the power series

$$\phi(z) \frac{dz}{dt} = \sum_0^\infty A_n t^n, \qquad (29.8)$$

it being assumed that $\phi(z)$ is regular in a neighbourhood of z_0. Evidently

$$A_n = \frac{1}{2\pi i} \int^{(0+)} \phi(z) \frac{dz}{dt} \frac{dt}{t^{n+1}}$$

$$= \frac{1}{2\pi i} \int^{(z_0+)} \frac{\phi(z)}{t^{n+1}} dz$$

$$= \frac{1}{2\pi i} \int^{(z_0+)} \frac{\phi(z)}{(w_0 - w)^{(n+1)/m}} dz,$$

with an appropriate branch of $(w_0 - w)^{1/m}$. Thus A_n is the residue of $\phi(z)/(w_0 - w)^{(n+1)/m}$ at z_0. In particular, $n c_n$ is the residue of $1/(w_0 - w)^{n/m}$. The actual determination of the coefficients may prove rather complicated in particular problems; it may only be possible to find the first few explicitly and not to get a general formula.

At z_0, we have $\qquad \dfrac{dz}{dt} = c_1 = \dfrac{A_0}{\phi(z_0)};$

and the phase of dz/dt at z_0 gives the direction of the corresponding path of steepest descent.

This account of Debye's method of steepest descents is necessarily merely a description, not a rigorous discussion; the fundamental idea will be made clearer by a consideration of particular examples. It should be noted that, in (29.4), we assumed ν large and positive. But it usually happens that the integrals obtained in this way converge when $|\nu|$ is large and $|\text{ph}\,\nu| < \alpha$, say; the method then gives an asymptotic expansion for large complex values of ν even though the path is then not a path of steepest descent.

30. The asymptotic expansion of $1/\Gamma(\nu)$

In this section we apply the method of steepest descents to Hankel's integral

$$\frac{1}{\Gamma(\nu)} = \frac{1}{2\pi i} \int_C e^t t^{-\nu} \, dt,$$

where t^ν means $e^{\nu \log t}$, the logarithm having its principal value. The path of integration starts at $\infty e^{-\pi i}$, goes round the origin once and ends at $\infty e^{\pi i}$. The formula is valid for all real or complex values of ν; but if ν is real and positive, the substitution $t = \nu z$ gives

$$\frac{1}{\Gamma(\nu)} = \frac{1}{2\pi i \nu^{\nu-1}} \int_C e^{\nu(z - \log z)} \, dz, \qquad (30.1)$$

with the same path as before. This integral converges uniformly on any compact set in $\mathscr{R}\nu > 0$; and since $1/\Gamma(\nu)$ is an integral function, equation (30.1) holds in this half-plane, provided that ν^ν means $e^{\nu \log \nu}$ where $\log \nu$ has its principal value.

There is only one saddle-point, namely $z = 1$. And when ν is positive, the paths of steepest descent are given by

$$z - \log z = 1 - \tau, \qquad (30.2)$$

where $\tau > 0$. If τ is small, z is approximately $1 \pm i \sqrt{(2\tau)}$, so that there are two directions of steepest descent from the saddle-point, in which $\text{ph}\,(z-1) = \pm \frac{1}{2}\pi$. The equation of the steepest paths is $\mathscr{I}(z - \log z) = 0$, which gives

$$y - \tan^{-1}\frac{y}{x} = 0.$$

This equation is satisfied by $y = 0$, which gives two directions of steepest ascent from the saddle-point, and by $x = y \cot y$. The upper and lower halves of this curve, which is symmetrical about the real axis, are the two paths of steepest descent. Since

$$\frac{dx}{dy} = \frac{\sin 2y - 2y}{2 \sin^2 y},$$

x decreases steadily as y increases from 0 to π, and tends to $-\infty$; and similarly for y decreasing from 0 to $-\pi$. The curve $x = y \cot y$, or rather the branch through $(1, 0)$, is thus asymptotic at $x = -\infty$ to $y = \pm \pi$. A simple application of Cauchy's theorem shows that it may be taken as the path of integration C.

If we put $z = 1 + Z$, $\tau = -\tfrac{1}{2}t^2$ in (30.2), we get the equation

$$Z - \log(1 + Z) = \tfrac{1}{2}t^2,$$

discussed in a slightly different notation in section 25. This equation defines a many-valued function $Z(t)$ of a complex variable t with branch-points at $t = \pm 2\sqrt{(n\pi)}\, e^{\pm \frac{1}{4}\pi i}$. The two solutions which vanish at $t = 0$ are

$$Z_1(t) = \sum_{1}^{\infty} a_n t^n, \quad Z_2(t) = \sum_{1}^{\infty} a_n(-t)^n,$$

the series being certainly convergent for $|t| < 2\sqrt{\pi}$, since the circle of convergence goes through the nearest singularity. The first five coefficients a_n were found before to be

$$a_1 = 1, \quad a_2 = \tfrac{1}{3}, \quad a_3 = \tfrac{1}{36}, \quad a_4 = -\tfrac{1}{270}, \quad a_5 = \tfrac{1}{4320}.$$

Replacing the original variable, we obtain two solutions of (30.2)

$$z_1 = 1 + \sum_{1}^{\infty} a_n i^n (2\tau)^{\frac{1}{2}n},$$

which gives the upper path of steepest descent when $\sqrt{(2\tau)} > 0$, and

$$z_2 = 1 + \sum_{1}^{\infty} a_n(-i)^n (2\tau)^{\frac{1}{2}n},$$

which gives the lower path. Hence

$$\frac{1}{\Gamma(\nu)} = \frac{e^\nu}{2\pi i \nu^{\nu-1}} \int_0^{\infty} e^{-\nu t} \left(\frac{dz_1}{d\tau} - \frac{dz_2}{d\tau} \right) d\tau. \tag{30.3}$$

Evidently $d(z_1 - z_2)/d\tau$ is an analytic function of the complex variable τ of the type specified in Watson's lemma. The only outstanding condition relates to its behaviour when τ is real and positive. Since $dz/d\tau = z/(1-z)$, we have

$$\frac{dz_1}{d\tau} - \frac{dz_2}{d\tau} = \frac{z_1}{1-z_1} - \frac{z_2}{1-z_2} = \frac{1}{z_2 - 1} - \frac{1}{z_1 - 1},$$

which is certainly bounded in $\tau \geqslant \tau_0$ for any positive value of τ_0. We may therefore substitute the power series for z_1 and z_2 in the integral and integrate term by term.

Before we do so, we observe that the integral (30.3) converges uniformly on any compact set in $|\nu| > 0$, $|\text{ph}\,\nu| < \frac{1}{2}\pi$, and that the representation of $1/\Gamma(\nu)$ is valid in this more extended region. The final result is that

$$\frac{1}{\Gamma(\nu)} \sim e^\nu \nu^{-\nu} \Big/ \sqrt{\left(\frac{\nu}{2\pi}\right)} \sum_1^\infty \frac{1\,.\,3\,.\,5\,.\,\ldots(2n+1)}{\nu^n} (-1)^n a_{2n+1},$$

when $|\text{ph}\,\nu| < \frac{1}{2}\pi$, or

$$\frac{1}{\Gamma(\nu)} \sim e^\nu \nu^{-\nu} \Big/ \sqrt{\left(\frac{\nu}{2\pi}\right)} \left\{ 1 - \frac{1}{12\nu} + \frac{1}{288\nu^2} - \ldots \right\}. \qquad (30.4)$$

That this result does, in fact, hold for $|\text{ph}\,\nu| \leqslant \pi - \epsilon < \pi$, can be shown by a suitable rotation of the path of integration in the τ-plane.

We could have deduced this result from the known asymptotic expansion of $\Gamma(\nu)$. It is given here as being probably the simplest example of Debye's method.

31. The Bessel function $J_\nu(a)$

A very full account of the theory of the asymptotic expansions of the Bessel functions will be found in Watson's *Theory of Bessel Functions*. A few simple cases will be discussed here as examples of the method of steepest descents, the simplest being that of $J_\nu(a)$ as $a \to +\infty$.

It is convenient to start with the Hankel function defined by

$$H_\nu^{(1)}(a) = \frac{1}{\pi i} \int_{-\infty}^{\infty + \pi i} e^{a \sinh z - \nu z} dz, \qquad (31.1)$$

when $|\mathrm{ph}\,a| < \tfrac{1}{2}\pi$. We assume that a is real and positive, and that the order ν is kept fixed. The parameter a plays the part played by ν in previous sections. The saddle-points are given by $\cosh z = 0$, and the only one which concerns us is $z = \tfrac{1}{2}\pi i$. The paths of steepest descent are then given by

$$\sinh z = i - \tau \quad (\tau > 0). \tag{31.2}$$

When τ is small, z is approximately equal to $\tfrac{1}{2}\pi i \pm e^{\tfrac{1}{4}\pi i}\,\sqrt{(2\tau)}$, so that the two directions of steepest descent make angles $\tfrac{1}{4}\pi$, $\tfrac{5}{4}\pi$ with the real axis, and the directions of steepest ascent (corresponding to $\tau < 0$) are in the perpendicular directions. The equation of the steepest paths is

$$\cosh x \sin y = 1;$$

and the level curves through the saddle-point, which are given by $\sinh x \cos y = 0$, are $x = 0$ and $y = \tfrac{1}{2}\pi$. In figure 3, the full curves are the level curves, the broken curves the steepest curves and the shaded areas the valleys. The steepest curves are asymptotic to the real axis and to $y = \pi$ as $x \to \pm\infty$. The two paths of steepest descent from the saddle-point go to $z = \infty + \pi i$ and to $z = -\infty$ and so may be taken as the path of integration.

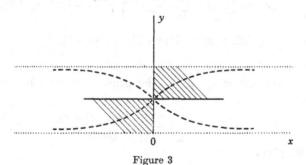

Figure 3

This path of integration is symmetrical about the saddle-point. If we put $z = \tfrac{1}{2}\pi i + \zeta$, we get

$$H_\nu^{(1)}(a) = \frac{1}{\pi i} \int_{-\infty-\frac{1}{2}\pi i}^{\infty+\frac{1}{2}\pi i} e^{ia\cosh\zeta - \nu\zeta - \frac{1}{2}\nu\pi i}\,d\zeta$$

$$= \frac{2}{\pi i} e^{-\frac{1}{2}\nu\pi i} \int_{0}^{\infty+\frac{1}{2}\pi i} e^{ia\cosh\zeta} \cosh\nu\zeta\,d\zeta,$$

F

where $$\cosh \zeta = 1 + i\tau. \qquad (31.3)$$
Hence
$$H_\nu^{(1)}(a) = \frac{2\,e^{ia - \frac{1}{2}\nu\pi i}}{\pi i} \int_0^\infty e^{-a\tau} \cosh \nu\zeta \frac{d\zeta}{d\tau} d\tau, \qquad (31.4)$$

to which we shall apply Watson's lemma. We note, however, that this formula holds, not only for $a > 0$, but also for

$$|\mathrm{ph}\,a| < \tfrac{1}{2}\pi.$$

The asymptotic expansion we shall obtain holds in this half-plane, though the path is not a path of steepest descent when a is complex.

From (31.3), we have

$$\sinh \tfrac{1}{2}\zeta = e^{\frac{1}{4}\pi i} \sqrt{(\tfrac{1}{2}\tau)},$$

where the square root is positive on the path of integration. If we regard τ as complex, $\sinh \tfrac{1}{2}\zeta$ is an analytic function of τ, regular save for a branch point at the origin. Also $\sinh \nu\zeta$ is an analytic function of $\sinh \tfrac{1}{2}\zeta$, regular when $|\sinh \tfrac{1}{2}\zeta| < 1$, and its Taylor series is

$$\sinh \nu\zeta = 2\nu \sinh \tfrac{1}{2}\zeta \left[1 + \sum_{r=1}^\infty \frac{(\tfrac{1}{2} - \nu)_r (\tfrac{1}{2} + \nu)_r}{(2r + 1)!} (-1)^r 2^{2r} \sinh^{2r} \tfrac{1}{2}\zeta \right],$$

where $$(\alpha)_r = \alpha(\alpha + 1)(\alpha + 2)\ldots(\alpha + r - 1).$$

Hence $\sinh \nu\zeta$ is an analytic function of τ, regular when $|\tau| < 2$ save for the branch-point at the origin, and can be written there as

$$\sinh \nu\zeta = 2\nu\, e^{\frac{1}{4}\pi i} \sqrt{(\tfrac{1}{2}\tau)} \left[1 + \sum_{r=1}^\infty \frac{(\tfrac{1}{2} - \nu)_r (\tfrac{1}{2} + \nu)_r}{(2r + 1)!} (-i\tau)^r \right].$$

And this gives

$$\cosh \nu\zeta \frac{d\zeta}{d\tau} = \frac{e^{\frac{1}{4}\pi i}}{\sqrt{(2\tau)}} \left[1 + \sum_{r=1}^\infty \frac{(\tfrac{1}{2} - \nu)_r (\tfrac{1}{2} + \nu)_r}{(2r)!} (-i\tau)^r \right],$$

also regular in $|\tau| < 2$ apart from the branch point.

Lastly, when τ is large and positive,

$$e^\zeta \sim 2i\tau, \qquad \frac{d\zeta}{d\tau} = \frac{i}{\sinh \zeta} \sim \frac{1}{\tau}.$$

Hence

$$\cosh \nu\zeta \frac{d\zeta}{d\tau} = \tfrac{1}{2}(e^{\nu\zeta}+e^{-\nu\zeta})\frac{d\zeta}{d\tau} \sim \frac{1}{2\tau}\{(2i\tau)^{\nu}+(2i\tau)^{-\nu}\},$$

and so there certainly exist constants K and τ_1 such that

$$\left|\cosh \nu\zeta \frac{d\zeta}{d\tau}\right| < K\,e^{\tau},$$

when $\tau > \tau_1$. As $\cosh \nu\zeta\, d\zeta/d\tau$ is evidently bounded for $1 \leqslant \tau \leqslant \tau_1$, all the conditions of Watson's lemma are satisfied. It follows that

$$H_{\nu}^{(1)}(a) \sim \left(\frac{2}{\pi a}\right)^{\!\frac{1}{2}} e^{i(a-\frac{1}{2}\nu\pi-\frac{1}{4}\pi)}\left[1+\sum_{r=1}^{\infty}\frac{(\tfrac{1}{2}-\nu)_r\,(\tfrac{1}{2}+\nu)_r}{r!\,(2ia)^r}\right]$$

as $|a| \to \infty$ in $|\mathrm{ph}\,a| \leqslant \tfrac{1}{2}\pi-\epsilon < \tfrac{1}{2}\pi$. The result, in fact, holds for $-\pi+\epsilon \leqslant |\mathrm{ph}\,a| \leqslant 2\pi-\epsilon < 2\pi$; the proof is somewhat tedious, but is on the same lines.

A similar argument shows that

$$H_{\nu}^{(2)}(a) \sim \left(\frac{2}{\pi a}\right)^{\!\frac{1}{2}} e^{-i(a-\frac{1}{2}\nu\pi-\frac{1}{4}\pi)}\left[1+\sum_{r=1}^{\infty}\frac{(\tfrac{1}{2}-\nu)_r\,(\tfrac{1}{2}+\nu)_r}{r!\,(-2ia)^r}\right],$$

as $|a| \to \infty$ in $-2\pi+\epsilon \leqslant \mathrm{ph}\,a \leqslant \pi-\epsilon < \pi$.

The asymptotic expansion

$$J_{\nu}(a) \sim \left(\frac{2}{\pi a}\right)^{\!\frac{1}{2}}\left\{\cos\left(a-\tfrac{1}{2}\nu\pi-\tfrac{1}{4}\pi\right)\sum_{r=0}^{\infty}\frac{(\tfrac{1}{2}-\nu)_{2r}(\tfrac{1}{2}+\nu)_{2r}}{(2r)!}\frac{(-1)^r}{(2a)^{2r}}\right.$$

$$\left.+\sin\left(a-\tfrac{1}{2}\nu\pi-\tfrac{1}{4}\pi\right)\sum_{r=0}^{\infty}\frac{(\tfrac{1}{2}-\nu)_{2r+1}(\tfrac{1}{2}+\nu)_{2r+1}}{(2r+1)!}\frac{(-1)^r}{(2a)^{2r+1}}\right\}$$

as $|a| \to \infty$ in $|\mathrm{ph}\,a| \leqslant \pi-\epsilon < \pi$ follows from these results and the formula $\qquad J_{\nu}(a) = \tfrac{1}{2}\{H_{\nu}^{(1)}(a)+H_{\nu}^{(2)}(a)\}.$

32. The Bessel function $J_{\nu}(\nu a)$

The problem of finding an asymptotic expansion of $J_{\nu}(a)$ when ν is large and a is fixed is of little interest; it can be readily shown from the power series for the Bessel function that

$$J_{\nu}(a) \sim \frac{e^{\nu}(\tfrac{1}{2}a)^{\nu}}{\sqrt{(2\pi\nu)}\,\nu^{\nu}}\left[1+\frac{a_1}{\nu}+\frac{a_2}{\nu^2}+\dots\right].$$

But the problem for $J_\nu(\nu a)$, when ν is large and a is fixed, is of very great interest, since the forms of the expansion are quite different when $a = 1$ and when a differs from unity.

We consider in this section the case $0 < a < 1$, and it is convenient to write $a = \operatorname{sech} \alpha$ where $\alpha > 0$, and $k = \nu a$. We assume that ν, and hence k, are large and positive. We then have

$$J_\nu(\nu a) = \frac{1}{2\pi i} \int_{\infty - \pi i}^{\infty + \pi i} e^{\nu a \sinh z - \nu z}\, dz$$

$$= \frac{1}{2\pi i} \int_{\infty - \pi i}^{\infty + \pi i} e^{k(\sinh z - z \cosh \alpha)}\, dz. \qquad (32.1)$$

The saddle-points are given by $\cosh z - \cosh \alpha = 0$, and so are $z = \pm \alpha + 2n\pi i$ where n is an integer. There are two saddle-points in the strip $|\mathscr{I}z| < \pi$ which concerns us, namely $z = \pm \alpha$, and only $z = \alpha$ is relevant.

The paths of steepest descent through $z = \alpha$ are given by

$$\sinh z - z \cosh \alpha = \sinh \alpha - \alpha \cosh \alpha - \tau, \qquad (32.2)$$

where $\tau > 0$. When τ is small, z is approximately equal to $\alpha \pm i \sqrt{(2\tau \operatorname{cosech} \alpha)}$, so that the tangents at the saddle-point to the two paths of steepest descent are parallel to the positive and negative directions of the imaginary axis. The steepest curves through $z = \alpha$ are given by $\cosh x \sin y - y \cosh \alpha = 0$. The paths of steepest ascent are $y = 0$, the positive and negative halves of the real axis; but the curves of steepest descent are the upper and lower halves of the branch of the curve

$$\cosh x = \frac{y \cosh \alpha}{\sin y}$$

through $z = \alpha$; this branch is symmetrical about the real axis. As y increases from 0 to π, x increases steadily from α to $+\infty$. Thus the two paths of steepest descent form an admissible path of integration in (32.1). The level curves through α are asymptotic to $y = \pm \frac{1}{2}\pi$ and terminate at the saddle-points $z = \alpha \pm 2\pi i$. (See figure 4).

If we write $\tau = t^2$ in (32.2), we have

$$\sinh z - z \cosh \alpha = \sinh \alpha - \alpha \cosh \alpha - t^2, \qquad (32.3)$$

where z is approximately $\alpha + it\sqrt{(2\operatorname{cosech}\alpha)}$ near the saddle-point, it being assumed that t is positive on the upper half and negative on the lower half of the path of integration. Equation (32.1) then becomes

$$J_\nu(\nu a) = \frac{e^{k(\sinh\alpha - \alpha\cosh\alpha)}}{2\pi i} \int_{-\infty}^{\infty} e^{-kt^2}\frac{dz}{dt}\,dt. \qquad (32.4)$$

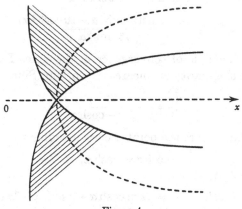

Figure 4

The equation (32.3) has two solutions, which are analytic functions of the complex variable t, regular in a neighbourhood of the origin. If we write

$$z = \alpha + c_1 t + c_2 t^2 + c_3 t^3 + \dots,$$

we require the solution for which $c_1 = i\sqrt{(2\operatorname{cosech}\alpha)}$. The easiest way of finding the coefficients, for which there does not appear to be a simple formula, is to differentiate (32.3) repeatedly and put $t = 0$, $z = \alpha$ at each step; the coefficient c_n is equal to the value of $z_n/n!$ at $t = 0$, where z_n denotes the nth derivative of z with respect to t. This process gives

$$(\cosh z - \cosh\alpha)z_1 = -2t,$$

$$(\cosh z - \cosh\alpha)z_2 + \sinh z\, z_1^2 = -2,$$

$$(\cosh z - \cosh\alpha)z_3 + 3\sinh z\, z_1 z_2 + \cosh z\, z_1^3 = 0,$$

$$(\cosh z - \cosh\alpha)z_4 + 4\sinh z\, z_1 z_3 + 3\sinh z\, z_2^2 + 6\cosh z\, z_1^2 z_2$$
$$+ \sinh z\, z_1^4 = 0$$

and so on. When $t = 0$, $z = \alpha$, the first equation is trivial and the second gives the known value of $z_1(0)$. From the third we get

$$z_2(0) = -\frac{\cosh\alpha}{3\sinh\alpha}\,z_1^2(0) = \frac{2\cosh\alpha}{3\sinh^2\alpha}.$$

Similarly, we obtain

$$z_3(0) = \frac{i\sqrt{2}}{\sqrt{\sinh\alpha}}\,\frac{3\sinh^2\alpha - 5\cosh^2\alpha}{6\sinh^3\alpha},$$

$$z_4(0) = \frac{16\cosh\alpha(9\sinh^2\alpha - 10\cosh^2\alpha)}{45\sinh^5\alpha},$$

and so on. The circle of convergence of the series $\Sigma c_n t^n$ passes through the singularity of z nearest the origin. Since

$$\frac{dz}{dt} = \frac{2t}{\cosh\alpha - \cosh z},$$

these singularities are the points t corresponding to

$$\cosh z = \cosh\alpha,$$

and are given by

$$t^2 = 2n\pi i \cosh\alpha, \quad t^2 = 2n\pi i \cosh\alpha + 2\sinh\alpha - 2\alpha\cosh\alpha,$$

where n assumes integer values ($t = 0$ being excluded). The series has therefore a non-zero radius of convergence.

Lastly, when t^2 is large and positive, it follows from (32.3) that e^z is approximately equal to $-2t^2$, and so dz/dt tends to zero as t tends to $+\infty$ or to $-\infty$. The conditions of the alternative form of Watson's lemma are thus satisfied. If we substitute in (32.4) the power series for z and integrate term by term, we find that

$$J_\nu(\nu a) \sim \frac{e^{k(\sinh\alpha - \alpha\cosh\alpha)}}{2\pi i}\sum_0^\infty (2n+1)c_{2n+1}\int_{-\infty}^\infty e^{-kt^2}t^{2n}\,dt$$

$$= \frac{e^{k(\sinh\alpha - \alpha\cosh\alpha)}}{2\pi i}\sum_0^\infty (2n+1)c_{2n+1}\frac{\Gamma(n+\tfrac12)}{k^{n+\frac12}},$$

as $k \to +\infty$, or, more generally, as $|k| \to \infty$ in

$$|\mathrm{ph}\,k| \leqslant \tfrac12\pi - \epsilon < \tfrac12\pi.$$

Restoring the original variables, we find that

$$J_\nu(\nu\,\mathrm{sech}\,\alpha) \sim \frac{e^{\nu(\tanh\alpha - \alpha)}}{2\pi i}\sum_0^\infty \frac{\Gamma(n+\tfrac12)}{(2n)!}z_{2n+1}(0)\frac{1}{(\nu\,\mathrm{sech}\,\alpha)^{n+\frac12}} \quad (32.5)$$

THE METHOD OF STEEPEST DESCENTS

for $\alpha > 0$, as $|\nu| \to \infty$ in $|\text{ph}\,\nu| \leqslant \tfrac{1}{2}\pi - \epsilon$ for any positive value of ϵ. The coefficients are real since $z_{2n+1}(0)$ is purely imaginary. The leading term of the expansion is

$$J_\nu(\nu \operatorname{sech} \alpha) \sim \frac{e^{\nu(\tanh \alpha - \alpha)}}{\sqrt{(2\pi\nu \tanh \alpha)}}.$$

33. The Bessel function $J_\nu(\nu)$

An example of a saddle-point of higher order is provided by the Bessel function $J_\nu(\nu)$ defined by

$$J_\nu(\nu) = \frac{1}{2\pi i} \int_{\infty - \pi i}^{\infty + \pi i} e^{\nu(\sinh z - z)}\,dz, \qquad (33.1)$$

when $\nu > 0$ or, more generally, when $\mathscr{R}\nu > 0$. The path of integration may consist of $y = -\pi$ from $x = +\infty$ to $x = 0$, $x = 0$ from $y = -\pi$ to $y = \pi$, and $y = \pi$ from $x = 0$ to $x = +\infty$.

The saddle-points, given by $\cosh z = 1$, are at $z = 2n\pi i$ when n is an integer or zero. Only the saddle-point at the origin concerns us here. When ν is positive, the paths of steepest descent from the origin are given by

$$\sinh z - z = -\tau \quad (\tau > 0). \qquad (33.2)$$

When τ is small, this is approximately $z^3 = -6\tau$, so that there are three directions of steepest descent from the origin, the directions in which the phases of z are $\pm \tfrac{1}{3}\pi$ or $-\pi$. There are also three directions of steepest ascent from the origin, the phases of z being $\pm \tfrac{2}{3}\pi$ or 0.

The curves of steepest ascent and descent from the origin are given by
$$\cosh x \sin y - y = 0,$$

and so are the real axis and the curve

$$\cosh x = \frac{y}{\sin y},$$

which has $y = \pm \pi$ as asymptotes. The level curves through the origin are the imaginary axis and the curve

$$\cos y = \frac{x}{\sinh x},$$

which has $y = \pm \frac{1}{2}\pi$ as asymptotes. In figure 5, the shaded areas are the valleys below the saddle-point, the unshaded areas the high ground and the broken lines the steepest curves from the saddle-point. We may evidently take as path of integration the broken line from $\infty - \pi i$ to the origin, followed by the steepest curve to $\infty + \pi i$. This gives

$$J_\nu(\nu) = \frac{1}{2\pi i} \left\{ \int_0^{\infty + \pi i} - \int_0^{\infty - \pi i} e^{\nu(\sinh z - z)}\, dz \right\},$$

where integration is along paths of steepest descent.

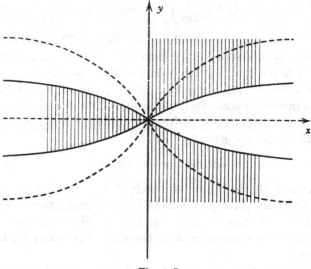

Figure 5

The three paths of steepest descent from the saddle-point have parametric equations obtained by finding the three solutions of

$$\sinh z - z = -\tau.$$

One solution $z = z_1(\tau)$ will give the path from 0 to $\infty + \pi i$, a second $z = z_2(\tau)$ the path from 0 to $\infty - \pi i$, and the third $z = z_3(\tau)$ the path from 0 to $-\infty$. We then have to see whether we can apply Watson's lemma to the resulting integral

$$J_\nu(\nu) = \frac{1}{2\pi i} \int_0^\infty e^{-\nu \tau} \left\{ \frac{dz_1}{d\tau} - \frac{dz_2}{d\tau} \right\} d\tau. \tag{33.3}$$

In order to apply the theory of analytic functions, we write $\tau = t^3$ and regard t for the moment as a complex variable. Since

$$\sinh z - z = z^3 \left\{ \frac{1}{3!} + \sum_1^\infty \frac{z^{2n}}{(2n+3)!} \right\},$$

the equation to be solved becomes

$$zf(z) = \omega t, \tag{33.4}$$

where ω is a complex cube root of -1; in fact $\omega = e^{\frac{1}{3}\pi i}$ for z_1, $\omega = e^{-\frac{1}{3}\pi i}$ for z_2, and $\omega = -1$ for z_3. The function $f(z)$ is the branch of

$$\left\{ \frac{1}{3!} + \sum_1^\infty \frac{z^{2n}}{(2n+3)!} \right\}^{\frac{1}{3}}$$

which is positive at the origin. By the theory of inverse functions, equation (33.4) has a unique solution

$$z = \sum_1^\infty a_n (\omega t)^n, \tag{33.5}$$

which vanishes at $t = 0$ and is regular in a neighbourhood of the origin. Since

$$\frac{dz}{dt} = \frac{-3t^2}{\cosh z - 1},$$

the singularities of any branch of z, regarded as a function of t, are the values of t corresponding to $z = 2n\pi i$, where n is a non-zero integer, and so satisfy $t^3 = -2n\pi i$. The series (33.5) is then certainly convergent for $|t| < (2\pi)^{\frac{1}{3}}$. If we put $t^3 = \tau$, (33.5) gives the three functions $z_r(\tau)$, each regular in $|\tau| < 2\pi$, apart from a branch point at $\tau = 0$.

Since z is an odd function of t, the coefficients a_{2n} in (33.5) are all zero. There does not appear to be a simple formula for the odd coefficients. These can be found successively by the method used in section 32 or by elementary algebra. The first three non-zero terms are

$$z = 6^{\frac{1}{3}}\omega t - \tfrac{1}{60}(6^{\frac{1}{3}}\omega t)^3 + \tfrac{1}{1400}(6^{\frac{1}{3}}\omega t)^5 + \dots. \tag{33.6}$$

The three paths of steepest descent from the origin are those on which $\tau^{\frac{1}{3}}$ is positive. When τ is large and positive, e^{z_1} and e^{z_2} are approximately equal to $-\tau$, and so

$$\frac{dz}{d\tau} = \frac{1}{1 - \cosh z}$$

is then bounded. The conditions of Watson's lemma are thus satisfied. Now

$$z_1 - z_2 = \sum_1^{\infty} a_n(\omega_1^n - \omega_2^n)\,\tau^{\frac{1}{3}n},$$

where $\omega_1 = e^{\frac{1}{3}\pi i}$, $\omega_2 = e^{-\frac{1}{3}\pi i}$, which gives

$$z_1 - z_2 = 2i \sum_1^{\infty} a_n \sin \tfrac{1}{3}n\pi\,.\,\tau^{\frac{1}{3}n}.$$

Therefore, when ν is large and positive,

$$J_\nu(\nu) \sim \frac{1}{\pi} \sum_1^{\infty} \tfrac{1}{3}na_n \sin \tfrac{1}{3}n\pi \int_0^{\infty} e^{-\nu\tau}\tau^{\frac{1}{3}n-1}\,d\tau,$$

or $$J_\nu(\nu) \sim \frac{1}{\pi} \sum_1^{\infty} a_n \sin \tfrac{1}{3}n\pi \frac{\Gamma(\tfrac{1}{3}n+1)}{\nu^{\frac{1}{3}n}}, \qquad (33.7)$$

all the coefficients a_{2n} being zero. The second non-vanishing term is $O(\nu^{-\frac{5}{3}})$.

This asymptotic expansion is not the limiting form of the expansion (32.5) of $J_\nu(\nu \operatorname{sech} \alpha)$; (33.7) is a series in inverse powers of $\nu^{\frac{1}{3}}$, whereas (32.5) is in inverse powers of $\nu^{\frac{1}{2}}$. The reason for this is that the two saddle-points $\pm \alpha$ associated with

$$\exp (\operatorname{sech} \alpha \sinh z - z)$$

coalesce into a single saddle-point of higher order at the origin when α tends to zero. In chapter 10, we show that it is possible to find an asymptotic expansion which holds uniformly in a neighbourhood of $\alpha = 0$, but it is not an asymptotic power series.

It should be noted that $J_\nu(\nu)$ is an analytic function of the complex variable ν, regular in the plane cut from $\nu = -\infty$ to $\nu = 0$, and that it can be represented by (33.1) or (33.3) when the real part of ν is positive. The asymptotic expansion (33.7) is therefore valid when $|\nu|$ is large and $|\operatorname{ph} \nu| \leqslant \tfrac{1}{2}\pi - \delta < \tfrac{1}{2}\pi$.

34. The error function

With one exception, all the integrals considered in this chapter have had infinite limits of integration. In the remaining case, the integral representation of the hypergeometric function used by Riemann, it happened that the path of steepest descent through the saddle-point passed through the ends of the finite path of integration, and so Debye's method would have given a complete asymptotic expansion. But, in general, for an integral of the form

$$\int_a^b e^{\nu f(z)}\phi(z)\,dz$$

with finite limits of integration, there is no path of steepest descents through a saddle-point and the ends of the range of integration, so that a somewhat different procedure is needed. An integral of this type can usually be written in the form

$$\int_a^{} e^{\nu f(z)}\phi(z)\,dz - \int_b^{} e^{\nu f(z)}\phi(z)\,dz$$

with infinite upper limit. We restrict our discussion to integrals with one limit infinite and illustrate the essential idea by means of the Error Function

$$\text{Erfc}\,\sigma = \int_\sigma^\infty e^{-t^2}\,dt. \tag{34.1}$$

Since
$$\text{Erfc}\,\sigma = \int_0^\infty e^{-t^2}\,dt - \int_0^\sigma e^{-t^2}\,dt$$
$$= \tfrac{1}{2}\sqrt{\pi} - \sum_0^\infty \frac{(-1)^n}{n!}\frac{\sigma^{2n+1}}{2n+1},$$

Erfc σ is an integral function of the complex variable σ, which satisfies
$$\text{Erfc}\,\sigma + \text{Erfc}\,(-\sigma) = \sqrt{\pi}. \tag{34.2}$$

It therefore suffices to consider the asymptotic behaviour in the half-plane $\mathscr{R}\sigma \geqslant 0$.

One way of proceeding is to suppose in the first instance that σ is positive and to make the substitution $t = \sigma\sqrt{(1+\tau)}$. This gives
$$\text{Erfc}\,\sigma = \tfrac{1}{2}\sigma e^{-\sigma^2}\int_0^\infty e^{-\sigma^2\tau}\frac{d\tau}{\sqrt{(1+\tau)}}.$$

This integral converges uniformly with respect to the complex variable σ on any compact set in $|\sigma| > 0$, $|\text{ph}\,\sigma| < \tfrac{1}{4}\pi$, so that the representation is valid in this angle. A straightforward application of Watson's lemma shows that

$$\text{Erfc}\,\sigma \sim \tfrac{1}{2}e^{-\sigma^2}\sum_0^\infty \frac{(-1)^n\,\Gamma(n+\tfrac{1}{2})}{\Gamma(\tfrac{1}{2})\,\sigma^{2n+1}} \tag{34.3}$$

as $|\sigma| \to \infty$ in $|\text{ph}\,\sigma| \leqslant \tfrac{1}{4}\pi - \delta < \tfrac{1}{4}\pi$.

To illustrate the point we wish to make, it is, however, necessary to consider complex values of σ from the start. If $\sigma = \sqrt{\nu}\,e^{\alpha i}$, where $\sqrt{\nu} > 0$ and $-\tfrac{1}{2}\pi < \alpha < \tfrac{1}{2}\pi$, the substitution $t = z\sqrt{\nu}$ in (34.1) gives

$$\text{Erfc}\,\sigma = \nu^{\frac{1}{2}}\int_{\omega}^{\infty} e^{-\nu z^2}\,dz, \qquad (34.4)$$

where we have written ω for $e^{\alpha i}$. There is one saddle-point, the origin, and the level curves through it are $y = \pm x$. The steepest

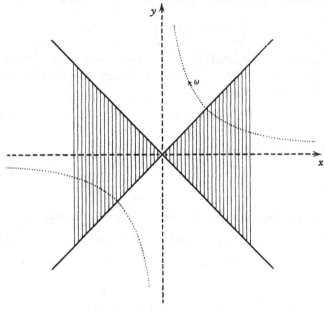

Figure 6

curves are the axes, and the valleys are the two shaded quadrants in the figure. The point ω lies on a path of steepest descent from a saddle-point only if $\alpha = 0$. But there is a unique path of steepest descent through ω, the curve on which

$$\mathscr{I}(-z^2) = \mathscr{I}(-\omega^2),$$

or
$$2xy = \sin 2\alpha.$$

There are three cases, $0 < \alpha < \tfrac{1}{2}\pi$, $\alpha = 0$, $-\tfrac{1}{2}\pi < \alpha < 0$. Figure 6 illustrates the first case. The broken lines are the steepest curves through the origin, the shaded areas the valleys below the origin, the dotted curve the steepest curve through ω. The following argument applies in all three cases.

No matter whether ω is in the valley below O or in the high ground above O, the hyperbolic path from ω to $+\infty$ runs steadily downhill. We may therefore put

$$-z^2 = -\omega^2 - \tau,$$

where τ increases steadily from 0 to $+\infty$ on this path. Since $\nu\omega^2 = \sigma^2$, equation (34.4) gives

$$\text{Erfc } \sigma = \tfrac{1}{2}\nu^{\frac{1}{2}} e^{-\sigma^2} \int_0^{\infty} e^{-\nu\tau}\left(1 + \frac{\tau}{\omega^2}\right)^{-\frac{1}{2}} \frac{d\tau}{\omega}.$$

When $|\text{ph } \sigma| \leqslant \tfrac{1}{2}\pi - \delta < \tfrac{1}{2}\pi$ and $\tau \geqslant 0$, the function $(1 + \tau\omega^{-2})^{-\frac{1}{2}}$ is bounded. For when $|\text{ph } \sigma| \leqslant \tfrac{1}{4}\pi$,

$$|\tau + \omega^2| = |z|^2 \geqslant |\omega^2| = 1,$$

and when $\tfrac{1}{4}\pi < |\text{ph } \sigma| \leqslant \tfrac{1}{2}\pi - \delta < \tfrac{1}{2}\pi$,

$$|\tau + \omega^2| = |z|^2 \geqslant |\mathscr{I}z^2| = \sin 2\alpha \geqslant \sin 2\delta.$$

We may therefore apply Watson's lemma to obtain

$$\text{Erfc } \sigma \sim \tfrac{1}{2}\nu^{\frac{1}{2}} e^{-\sigma^2} \sum_0^{\infty} \int_0^{\infty} e^{-\nu\tau}\frac{\Gamma(n+\tfrac{1}{2})}{\Gamma(\tfrac{1}{2})\,n!}\frac{(-1)^n\,\tau^n}{\omega^{2n+1}}\,d\tau$$

$$= \tfrac{1}{2}\nu^{\frac{1}{2}} e^{-\sigma^2} \sum_0^{\infty} \frac{\Gamma(n+\tfrac{1}{2})}{\Gamma(\tfrac{1}{2})}\frac{(-1)^n}{\nu^{n+1}\omega^{2n+1}},$$

or, restoring the original variable,

$$\text{Erfc } \sigma \sim \tfrac{1}{2}e^{-\sigma^2} \sum_0^{\infty} \frac{(-1)^n\,\Gamma(n+\tfrac{1}{2})}{\Gamma(\tfrac{1}{2})\,\sigma^{2n+1}},$$

which is (34.3), proved now for $|\text{ph } \sigma| \leqslant \tfrac{1}{2}\pi - \delta$.

If $\alpha = \pm\tfrac{1}{2}\pi$, the method fails, since the path degenerates into the imaginary axis from $\pm i$ to 0, followed by the real axis from 0 to $+\infty$. If $\alpha = \tfrac{1}{2}\pi$, this path gives

$$\text{Erfc } \sigma = -\sigma \int_0^1 e^{\nu y^2}\,dy + \nu^{\frac{1}{2}} \int_0^{\infty} e^{-\nu x^2}\,dx.$$

If we put $y^2 = 1 - \tau$, we obtain

$$\text{Erfc } \sigma = -\tfrac{1}{2}\sigma e^{-\sigma^2} \int_0^1 e^{-\nu\tau}\frac{d\tau}{\sqrt{(1-\tau)}} + \tfrac{1}{2}\sqrt{\pi},$$

and the methods we have developed so far will not give the asymptotic expansion of the first integral on the right-hand side. One way out of the difficulty is to start again from (34.4), namely

$$\text{Erfc}\,\sigma = \nu^{\frac{1}{2}} \int_{i}^{\infty} e^{-\nu z^2}\,dz, \qquad (34.5)$$

when $\alpha = \frac{1}{2}\pi$, and integrate along the level curve through i. On this level curve, $\mathscr{R}z^2$ is constant, so that

$$x^2 - y^2 = -1.$$

This hyperbola has $y = x$ as asymptote, and a modification of the Jordan's lemma argument shows that it is an admissible path. The parametric equation of this level curve is $z^2 = -1 + i\tau$, where τ increases from 0 to $+\infty$. Since $z = i\sqrt{(1-i\tau)}$, equation (34.5) gives

$$\text{Erfc}\,\sigma = \tfrac{1}{2}\nu^{\frac{1}{2}} e^{-\sigma^2} \int_{0}^{\infty} \frac{e^{-i\nu\tau}}{(1-i\tau)^{\frac{1}{2}}}\,d\tau,$$

whence, by integration by parts,

$$\text{Erfc}\,\sigma = \frac{e^{-\sigma^2}}{2i\nu^{\frac{1}{2}}} \sum_{0}^{N-1} \frac{\Gamma(n+\frac{1}{2})}{\Gamma(\frac{1}{2})\,\nu^n} + \frac{e^{-\sigma^2}}{2\nu^{\frac{1}{2}}} \frac{\Gamma(N+\frac{1}{2})}{\Gamma(\frac{1}{2})\,\nu^{N-1}} I_N,$$

where

$$I_N = \int_{0}^{\infty} \frac{e^{-i\nu\tau}}{(1-i\tau)^{N+\frac{1}{2}}}\,d\tau.$$

But

$$|I_N| \leqslant \int_{0}^{\infty} \frac{d\tau}{(1+\tau^2)^{\frac{1}{2}N+\frac{1}{4}}} = O(1)$$

as $\nu \to \infty$. Hence

$$\frac{e^{-\sigma^2}}{2i\nu^{\frac{1}{2}}} \frac{\Gamma(N-\frac{1}{2})}{\Gamma(\frac{1}{2})\,\nu^{N-1}} + \frac{e^{-\sigma^2}}{2\nu^{\frac{1}{2}}} \frac{\Gamma(N+\frac{1}{2})}{\Gamma(\frac{1}{2})\,\nu^{N-1}} I_N = O\!\left(\frac{e^{-\sigma^2}}{2\nu^{\frac{1}{2}}} \frac{\Gamma(N+\frac{1}{2})}{\Gamma(\frac{1}{2})\,\nu^{N-1}}\right).$$

This shows that

$$\text{Erfc}\,\sigma \sim \frac{e^{-\sigma^2}}{2i\nu^{\frac{1}{2}}} \sum_{0}^{\infty} \frac{\Gamma(n+\frac{1}{2})}{\Gamma(\frac{1}{2})\,\nu^n},$$

as $\nu \to \infty$, since the error after the term with $n = N-2$ is of the same order of magnitude as the term with $n = N-1$.

Restoring the original variable, we obtain

$$\text{Erfc}\,\sigma \sim \tfrac{1}{2} e^{-\sigma^2} \sum_{0}^{\infty} \frac{\Gamma(n+\frac{1}{2})\,(-1)^n}{\Gamma(\frac{1}{2})\,\sigma^{2n+1}}, \qquad (34.6)$$

as $|\sigma| \to \infty$, with $\mathrm{ph}\,\sigma = \tfrac{1}{2}\pi$. If we use the identity (34.2) we find that

$$\mathrm{Erfc}\,\sigma \sim \sqrt{\pi} + \tfrac{1}{2}e^{-\sigma^2} \sum_0^\infty \frac{\Gamma(n+\tfrac{1}{2})(-1)^n}{\Gamma(\tfrac{1}{2})\,\sigma^{2n+1}},$$

as $|\sigma| \to \infty$, with $\mathrm{ph}\,\sigma = -\tfrac{1}{2}\pi$. But $\sqrt{\pi}$ is very small compared with any term of the series, and so (34.6) holds with $\mathrm{ph}\,\sigma = \pm\tfrac{1}{2}\pi$. Thus the expansion (34.3) is now seen to hold when $|\mathrm{ph}\,\sigma| \leqslant \tfrac{1}{2}\pi$.

Lastly, using (34.2) again, we find that

$$\mathrm{Erfc}\,\sigma \sim \sqrt{\pi} + \tfrac{1}{2}e^{-\sigma^2} \sum_0^\infty \frac{\Gamma(n+\tfrac{1}{2})(-1)^n}{\Gamma(\tfrac{1}{2})\,\sigma^{2n+1}},$$

as $|\sigma| \to \infty$ in $\mathscr{R}\sigma < 0$.

It is possible to discuss in a similar way the asymptotic behaviour of

$$\int_r^s e^{i(x^m - \sigma x)}\,dx, \tag{34.7}$$

when σ, r, s are large. Brillouin [2] treated the cases $m = 3$ and 4 for real values of σ, r, s: he pointed out that, whilst we may write this integral in the form

$$\int^s - \int^r,$$

with the same infinite lower limit in each, trouble may arise if we substitute for these integrals asymptotic expansions valid for large r and s in the case when $r - s$ is small. Burwell [3] has discussed the integral (34.7) when m is any positive integer and σ, r and s are complex.

35. Another integral with finite limits

An interesting example of an integral with finite limits is

$$F(\sigma) = \int_0^1 e^{\sigma t^3}\,dt. \tag{35.1}$$

$F(\sigma)$ is an integral function of the complex variable σ with Taylor series

$$\sum_0^\infty \frac{\sigma^n}{n!\,(3n+1)}.$$

When $\mathscr{R}\sigma > 0$, we can write

$$F(\sigma) = \int_{-\infty}^{1} e^{\sigma t^3}\, dt - \int_{-\infty}^{0} e^{\sigma t^3}\, dt$$

$$= \tfrac{1}{3}e^{\sigma}\int_{0}^{\infty} e^{-\sigma\tau}(1-\tau)^{-\frac{2}{3}}\, d\tau - \frac{1}{3}\int_{0}^{\infty} e^{-\sigma\tau}\,\tau^{-\frac{2}{3}}\, d\tau.$$

The second integral can be evaluated in terms of Gamma functions; to the first, Watson's lemma can be applied. The result is that

$$F(\sigma) \sim \tfrac{1}{3}e^{\sigma}\sum_{0}^{\infty} \frac{\Gamma(n+\frac{2}{3})}{\Gamma(\frac{2}{3})}\,\frac{1}{\sigma^{n+1}} - \frac{\Gamma(\frac{4}{3})}{\sigma^{\frac{1}{3}}} \tag{35.2}$$

as $|\sigma| \to \infty$ in $|\mathrm{ph}\,\sigma| \leqslant \tfrac{1}{2}\pi - \delta < \tfrac{1}{2}\pi$. But the second term is small compared with any term in the infinite series.

When $\mathscr{R}\sigma < 0$, we write $\sigma = -\nu$ where $\mathscr{R}\nu > 0$. Then

$$F(-\nu) = \int_{0}^{\infty} e^{-\nu t^3}\, dt - \int_{1}^{\infty} e^{-\nu t^3}\, d\tau$$

$$= \frac{1}{3}\int_{0}^{\infty} e^{-\nu\tau}\,\tau^{-\frac{2}{3}}\, d\tau - \tfrac{1}{3}e^{-\nu}\int_{0}^{\infty} e^{-\nu\tau}(1+\tau)^{-\frac{2}{3}}\, d\tau$$

$$\sim \frac{\Gamma(\frac{4}{3})}{\nu^{\frac{1}{3}}} + \tfrac{1}{3}e^{-\nu}\sum_{0}^{\infty} \frac{\Gamma(n+\frac{2}{3})}{\Gamma(\frac{2}{3})}\,\frac{(-1)^{n+1}}{\nu^{n+1}},$$

as $|\nu| \to \infty$ in $|\mathrm{ph}\,\nu| \leqslant \tfrac{1}{2}\pi - \delta < \tfrac{1}{2}\pi$. Restoring the original variable,

$$F(\sigma) \sim \tfrac{1}{3}e^{\sigma}\sum_{0}^{\infty} \frac{\Gamma(n+\frac{2}{3})}{\Gamma(\frac{2}{3})}\,\frac{1}{\sigma^{n+1}} + \frac{\Gamma(\frac{4}{3})}{(-\sigma)^{\frac{1}{3}}}, \tag{35.3}$$

when $|\mathrm{ph}\,(-\sigma)| \leqslant \tfrac{1}{2}\pi - \delta$. This differs from (35.2) in that the term in $\sigma^{-\frac{1}{3}}$ is now much larger than any term of the infinite series.

The preceding argument gives the asymptotic expansion over the whole plane, except in two small angles enclosing the positive and negative halves of the imaginary axis. To complete the discussion, we use the method of steepest descents.

Let us write $\sigma = \nu i$ where $\mathscr{R}\nu > 0$. We now have

$$F(i\nu) = \int_{0}^{1} e^{i\nu z^3}\, dz. \tag{35.4}$$

There is one saddle-point, the origin. If $\nu > 0$, the level curves through the saddle-point are $y = 0$ and $y = \pm \sqrt{3}x$, and the steepest curves are $x = 0$ and $x = \pm \sqrt{3}y$. Three valleys meet at the saddle-point, as shown in figure 7. The points 0 and 1 do not lie on a steepest path through a saddle-point—they both lie on the same level curve.

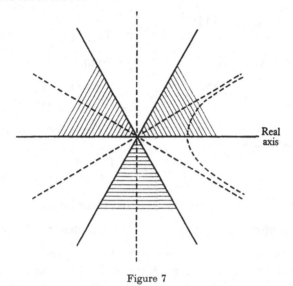

Figure 7

The steepest curve through $z = 1$ is given by $\mathscr{I}(iz^3) = \mathscr{I}(i)$. This is a cubic curve $x^3 - 3xy^2 = 1$, and the branch of it through $z = 1$ has $x = \sqrt{3}y$ as an asymptote. The upper half of the branch is a path of steepest descent from 1 to $\infty e^{\frac{2}{3}\pi i}$. Accordingly we shall integrate first along the path of steepest descent $\mathrm{ph}\, z = \frac{1}{6}\pi$ from the origin to $\infty e^{\frac{1}{6}\pi i}$ and then from $\infty e^{\frac{2}{3}\pi i}$ to 1 along the upper half of the branch of the cubic. On the path from 0 to $\infty e^{\frac{1}{6}\pi i}$, $z = \tau^{\frac{1}{3}} e^{\frac{1}{6}\pi i}$ where $\tau > 0$; and on the steepest path from 1 to $\infty e^{\frac{2}{3}\pi i}$,

$$iz^3 = i - \tau,$$

and so

$$z = (1 + i\tau)^{\frac{1}{3}}.$$

Making these substitutions, we find that, when $\nu > 0$,

$$F(i\nu) = \tfrac{1}{3}e^{\frac{1}{6}\pi i}\int_0^\infty e^{-\nu\tau}\tau^{-\frac{2}{3}}\,d\tau - \tfrac{1}{3}e^{i\nu}\int_0^\infty e^{-\nu\tau}(1 + i\tau)^{-\frac{2}{3}}\,i\,d\tau,$$

G

and, by analytical continuation, this formula holds when $|\text{ph}\,\nu| < \frac{1}{2}\pi$. The first integral can be evaluated at once, and Watson's lemma gives an asymptotic expansion of the second integral. It follows that

$$F(i\nu) \sim \frac{\Gamma(\frac{4}{3})\,e^{\frac{1}{6}\pi i}}{\nu^{\frac{1}{3}}} + \tfrac{1}{3}e^{i\nu}\sum_{0}^{\infty}\frac{\Gamma(n+\frac{2}{3})}{\Gamma(\frac{2}{3})}\frac{1}{(i\nu)^{n+1}},$$

when $|\nu| \to \infty$ in $|\text{ph}\,\nu| \leqslant \frac{1}{2}\pi - \delta < \frac{1}{2}\pi$. Similarly,

$$F(-i\nu) \sim \frac{\Gamma(\frac{4}{3})\,e^{-\frac{1}{6}\pi i}}{\nu^{\frac{1}{3}}} + \tfrac{1}{3}e^{-i\nu}\sum_{0}^{\infty}\frac{\Gamma(n+\frac{2}{3})}{\Gamma(\frac{2}{3})}\frac{1}{(-i\nu)^{n+1}},$$

when $|\nu| \to \infty$ in $|\text{ph}\,\nu| \leqslant \frac{1}{2}\pi - \delta < \frac{1}{2}\pi$. These can be combined into one formula

$$F(\sigma) \sim \frac{\Gamma(\frac{4}{3})\,e^{\pm\frac{1}{3}\pi i}}{\sigma^{\frac{1}{3}}} + \tfrac{1}{3}e^{\sigma}\sum_{0}^{\infty}\frac{\Gamma(n+\frac{2}{3})}{\Gamma(\frac{2}{3})}\frac{1}{\sigma^{\nu+1}}, \qquad (35.5)$$

where the upper or lower sign is taken according as

$$\delta \leqslant \text{ph}\,\sigma \leqslant \pi - \delta < \pi \quad \text{or} \quad -\pi + \delta \leqslant \text{ph}\,\sigma \leqslant -\delta < 0.$$

We now have four different asymptotic expansions for $F(\sigma)$, valid in four overlapping angles, yet the formulae are not inconsistent if regard is had to the phase of σ, an example of the Stokes phenomenon.

CHAPTER 8

THE SADDLE-POINT METHOD

36. Explanation of the method

We saw in chapter 7 that, if $w(z)$ and $\phi(z)$ are analytic functions, regular in a certain domain of the complex plane, it is often possible to find a complete asymptotic expansion of the integral

$$\int e^{\nu w(z)} \phi(z)\, dz \tag{36.1}$$

by a suitable deformation of the path of integration. In most cases we make the deformed path pass through one or more saddle-points at which $w'(z)$ vanishes, and construct the path out of paths of steepest descent. If we write $w = u + iv$, where u and v are real, v is constant on a path of steepest descent, and the dominant part of the asymptotic expansion arises from the part of the path near the highest saddle-point, the point where u takes its greatest value.

If z_0 is a saddle-point, the paths of steepest descent from z_0 are given by
$$w(z) = w(z_0) - \tau,$$

where τ is positive. The method of steepest descents is often cumbersome because of the difficulty in expressing z in terms of τ. We now describe a simpler procedure which we call the *saddle-point method*.

Let us suppose that it is possible to deform the path of integration in (36.1) into a path which goes through one or more saddle-points and which lies in the valleys below these saddle-points, without altering the value of the integral. If z_0 is the highest of these saddle-points, that is, z_0 is the saddle-point at which u takes its greatest value, the neighbourhood of z_0 provides the dominant part of the integral as $\nu \to +\infty$. There is no loss of generality in assuming that $\nu \to +\infty$ rather than $|\nu| \to \infty$; for if $\nu \to \infty e^{i\gamma}$, we can write $\nu w(z) = \nu_1 e^{i\gamma} w(z) = \nu_1 w_1(z)$, where $\nu_1 \to +\infty$.

Evidently, if there are several saddle-points at the same height, each makes a contribution of the same order of magnitude, and we can consider each separately. There is considerable freedom in the choice of the path of integration. It does not need to consist of paths of steepest descent, but there are restrictions on its direction at the saddle-point.

In the simplest case, the saddle-point z_0 is a simple zero of $w'(z)$. Near z_0, $w(z)$ can be expanded as a convergent power series

$$w(z) = w(z_0) + a_2(z - z_0)^2 + a_3(z - z_0)^3 + \ldots, \qquad (36.2)$$

where $a_2 = \frac{1}{2}w''(z_0)$ and so on. We choose the path of integration near z_0 to be a straight line on which the second term in the series (36.2) is real and negative. The direction of this line may be called the *critical direction at* z_0; it is the tangent at z_0 to the two directions of steepest descent from z_0.

Suppose that the power-series (36.2) is convergent in

$$|z - z_0| \leqslant R$$

and that M is the maximum of $|w(z)|$ on $|z - z_0| = R$. Then

$$|a_n| \leqslant \frac{M}{R^n},$$

by Cauchy's inequality. Hence if

$$w(z) = w(z_0) + a_2(z - z_0)^2 + F(z), \qquad (36.3)$$

we have
$$|F(z)| \leqslant M \left\{ \frac{|z - z_0|^3}{R^3} + \frac{|z - z_0|^4}{R^4} + \ldots \right\}$$

$$= \frac{M|z - z_0|^3}{R^2\{R - |z - z_0|\}}.$$

But if ϵ is positive, we can choose ν so large that $\nu^{-\epsilon} \leqslant \frac{1}{2}R$. It follows that, if $|z - z_0| \leqslant \nu^{-\epsilon}$,

$$|F(z)| \leqslant \frac{2M}{R^3}\nu^{-3\epsilon}.$$

The three terms on the right-hand side of (36.3) are therefore of the orders $O(1)$, $O(\nu^{-2\epsilon})$ and $O(\nu^{-3\epsilon})$ respectively, and so

$$e^{\nu w(z)} = \exp\{\nu w(z_0) + \nu a_2(z - z_0)^2\}\{1 + O(\nu^{1-3\epsilon})\}.$$

If we take $\epsilon > \frac{1}{3}$, the order term is very small when ν is large. Similarly,
$$\phi(z) = \phi(z_0) + O(\nu^{-\epsilon}) = \phi(z_0) + o(\nu^{1-3\epsilon}),$$

if $\epsilon < \frac{1}{2}$. We assume in what follows that ϵ has been chosen so that $\frac{1}{3} < \epsilon < \frac{1}{2}$. The contribution of the neighbourhood $|z - z_0| \leqslant \nu^{-\epsilon}$ of the saddle-point z_0 is therefore

$$\phi(z_0)\, e^{\nu w(z_0)} \int e^{\nu a_2 (z-z_0)^2}\, dz \{1 + O(\nu^{1-3\epsilon})\}. \qquad (36.4)$$

If we write $a_2 = A\, e^{\alpha i}$, where $A > 0$, and $z = z_0 + r\, e^{\theta i}$, then

$$a_2 (z - z_0)^2 = A r^2 e^{(\alpha + 2\theta)\, i}, \qquad (36.5)$$

which is real and negative when $\theta = \pm \frac{1}{2}\pi - \frac{1}{2}\alpha$. This gives two opposite directions corresponding to the two directions of steepest descent at z_0. We take the upper sign and let r vary from $-\eta$ to η, where $\eta = \nu^{-\epsilon}$. The expression (36.4) then becomes

$$\phi(z_0)\, e^{\nu w(z_0)} \int_{-\eta}^{\eta} e^{-A\nu r^2 + \frac{1}{2}(\pi - \alpha)\, i}\, dr,$$

omitting the factor $1 + O(\nu^{1-3\epsilon})$. If we put $A\nu r^2 = u^2$, we get

$$\phi(z_0)\, e^{\nu w(z_0) + \frac{1}{2}(\pi - \alpha)\, i} \frac{1}{\sqrt{(A\nu)}} \int_{-\omega}^{\omega} e^{-u^2}\, du, \qquad (36.6)$$

where $\qquad \omega = \sqrt{(A\nu\eta^2)} = \sqrt{(A\nu^{1-2\epsilon})}$.

Since $\epsilon < \frac{1}{2}$, ω tends to $+\infty$ with ν. But, as $\nu \to +\infty$,

$$\int_{\omega}^{\infty} e^{-u^2}\, du = O\!\left(\frac{e^{-\omega^2}}{\omega}\right) = O(\nu^{\epsilon - \frac{1}{2}} e^{-A\nu^{1-2\epsilon}}) = o(\nu^{1-3\epsilon}),$$

and similarly for the integral from $-\infty$ to $-\omega$. We may therefore replace the limits of integration in (36.6) by $\pm\infty$ without altering the factor $1 + O(\nu^{1-3\epsilon})$, which should occur but was omitted for brevity.

We have thus shown that the contribution (36.6) of the neighbourhood of the saddle-point z_0 is

$$\phi(z_0)\, e^{\nu w(z_0)} \left(\frac{-1}{A\nu\, e^{\alpha i}}\right)^{\frac{1}{2}} \int_{-\infty}^{\infty} e^{-u^2}\, du = \phi(z_0)\, e^{\nu w(z_0)} \left(\frac{-\pi}{A\nu\, e^{\alpha i}}\right)^{\frac{1}{2}}$$

$$= \phi(z_0)\, e^{\nu w(z_0)} \left(\frac{-2\pi}{\nu w''(z_0)}\right)^{\frac{1}{2}}. \qquad (36.7)$$

This is the required asymptotic approximation to the integral (36.1) when there is only one saddle-point, a simple zero of $w'(z)$. If there are several saddle-points at the same height, the approximation will consist of several terms like (36.7).

Sometimes it is simpler not to take the path at the saddle-point in the critical direction, but in a direction making an angle less than $\frac{1}{4}\pi$ with the critical direction. It is evident by (36.5) that this makes $a_2(z-z_0)^2$ complex with negative real part, and the analysis goes through with little change. But if we take a direction making an angle of $\frac{1}{4}\pi$ with the critical direction, the real part of $a_2(z-z_0)^2$ is zero and we are then, in effect, using the method of stationary phase.

Lastly, it will be recalled that we chose a path of integration lying in the valleys below the saddle-point. If we wish to make an estimate of the error involved in using the asymptotic approximation (36.7), it is usually simplest to choose a path on which the real part of $\nu w(z)$ decreases steadily.

We have considered here only the case when the highest saddle-point is a simple zero of $w'(z)$. If it is a zero of higher order, the argument becomes more complicated but not essentially different.

37. The Legendre polynomials

It follows from the formula of Rodrigues that the Legendre polynomial $P_n(\mu)$ can be represented by an integral

$$P_n(\mu) = \frac{1}{2^{n+1}\pi i} \int_C \frac{(z^2-1)^n}{(z-\mu)^{n+1}}\,dz, \qquad (37.1)$$

where C is a simple closed contour surrounding the point $z = \mu$. We use the saddle-point method to find an asymptotic approximation for $P_n(\mu)$ when the positive integer n is large and μ is real. There are two cases to consider, namely $-1 < \mu < 1$ and $\mu > 1$. We need not consider either the case $\mu < -1$, since

$$P_n(-\mu) = (-1)^n P_n(\mu);$$

or the cases $\mu = \pm 1$, since $P_n(1) = 1$.

The integral (37.1) is of the form

$$P_n(\mu) = \frac{1}{2^{n+1}\pi i} \int_C e^{nw(z)} \phi(z)\, dz, \qquad (37.2)$$

where $\qquad w(z) = \log(z^2 - 1) - \log(z - \mu),$

the logarithms having their principal values, and

$$\phi(z) = \frac{1}{z - \mu}.$$

Since $\qquad w'(z) = \dfrac{z^2 - 2\mu z + 1}{(z^2 - 1)(z - \mu)},$

there are two saddle-points, $z = \mu \pm \sqrt{(\mu^2 - 1)}.$

If $\mu = \cos\theta$, where $0 < \theta < \pi$, the two saddle-points are at $z = e^{\pm\theta i}$. Since
$$|\exp w(e^{\pm\theta i})| = 1,$$

the two saddle-points are equally important. But we do not need to consider them separately; for if we take C to be the circle $|z| = 1$, it is evident that

$$P_n(\mu) = \frac{1}{2^n\pi} \mathscr{I}I,$$

where $\qquad I = \displaystyle\int_\Gamma e^{nw(z)} \phi(z)\, dz, \qquad (37.3)$

Γ being the semicircle $|z| = 1$, $0 \leqslant \mathrm{ph}\, z \leqslant \pi$.

Near $z = e^{\theta i}$, we have

$$w(z) = \theta i + \log 2 - \tfrac{1}{2} i\, e^{-\theta i} \operatorname{cosec}\theta (z - e^{\theta i})^2 + \dots .$$

If we put $z = e^{\theta i} + r e^{\phi i}$,

$$w(z) = \theta i + \log 2 - i r^2 \operatorname{cosec}\theta\, e^{(2\phi - \theta) i} + \dots .$$

In the critical direction the coefficient of r^2 is real and negative, so that
$$\phi = \tfrac{1}{2}\theta + \tfrac{1}{4}\pi \pm \tfrac{1}{2}\pi.$$

Hence the critical direction is that of the bisector of the acute angle between the tangent to Γ at $e^{\theta i}$ and the real axis. As the critical direction at $e^{\theta i}$ always makes an angle less than $\tfrac{1}{4}\pi$ with the tangent to the circle, we can use the semicircle Γ as path of integration. And, although we integrated along a small straight segment in section 36, there is no need to do so—a small arc about the saddle-point does just as well; in fact we map the semicircle Γ on to the real line by writing $z = e^{it}$.

Consider first the neighbourhood $-\eta \leqslant t - \theta \leqslant \eta$ of the saddle-point, where $\eta = n^{-\epsilon}$ with $\frac{1}{3} < \epsilon < \frac{1}{2}$. On this arc,

$$nw(z) = n\theta i + \log 2^n + \tfrac{1}{2}ni\,e^{\theta i}\operatorname{cosec}\theta\,(t-\theta)^2 + O(n^{1-3\epsilon}),$$

and

$$\phi(z)\frac{dz}{dt} = \frac{i\,e^{it}}{e^{it}-\cos\theta} = \frac{e^{i\theta}}{\sin\theta}\{1+O(n^{-\epsilon})\} = \frac{e^{i\theta}}{\sin\theta}\{1+o(n^{1-3\epsilon})\},$$

since $\epsilon < \frac{1}{2}$. Therefore the contribution to I of the neighbourhood of the saddle-point is

$$I_1 = \frac{2^n\,e^{(n+1)\theta i}}{\sin\theta}\int_{\theta-\eta}^{\theta+\eta}\exp\{\tfrac{1}{2}in\,e^{\theta i}\operatorname{cosec}\theta\,(t-\theta)^2\}\,dt\,.\,\{1+O(n^{1-3\epsilon})\}$$

$$= \frac{2^n\,e^{(n+1)\theta i}}{\sin\theta}\left(\frac{2}{n}\right)^{\frac{1}{2}}\int_{-\omega}^{\omega}\exp\{-(1-i\cot\theta)\,u^2\}\,du\,.\,\{1+O(n^{1-3\epsilon})\},$$

where $\omega = \sqrt{(\frac{1}{2}n^{1-2\epsilon})}$, which tends to infinity with n. But

$$\left|\int_{\omega}^{\infty}\exp\{-(1-i\cot\theta)\,u^2\}\,du\right| \leqslant \int_{\omega}^{\infty}e^{-u^2}\,du < \frac{e^{-\omega^2}}{\omega}$$

$$= \frac{\exp(-\tfrac{1}{2}n^{1-2\epsilon})}{\sqrt{(\tfrac{1}{2}n^{1-2\epsilon})}} = o(n^{1-3\epsilon}),$$

and similarly for the integral from $-\infty$ to $-\omega$. Hence

$$I_1 = \frac{2^n\,e^{(n+1)\theta i}}{\sin\theta}\left(\frac{2}{n}\right)^{\frac{1}{2}}\int_{-\infty}^{\infty}\exp\{-(1-i\cot\theta)\,u^2\}\,du\,.\,\{1+O(n^{1-3\epsilon})\}$$

$$= \frac{2^n\,e^{(n+1)\theta i}}{\sin\theta}\left(\frac{2\pi}{n(1-i\cot\theta)}\right)^{\frac{1}{2}}\{1+O(n^{1-3\epsilon})\}$$

$$= 2^n\,e^{(n+\frac{1}{2})\theta i+\frac{1}{4}\pi i}\left(\frac{2\pi}{n\sin\theta}\right)^{\frac{1}{2}}\{1+O(n^{1-3\epsilon})\}.$$

It remains to consider the contribution of the other arcs of Γ. On Γ

$$|e^{w(z)}| = \left|\frac{e^{2it}-1}{e^{it}-\cos\theta}\right| = 2\left\{1+\frac{(\cos t-\cos\theta)^2}{\sin^2 t}\right\}^{-\frac{1}{2}},$$

which increases as t increases from 0 to θ and decreases as t increases from θ to π. Also

$$\left| \phi(z) \frac{dz}{dt} \right| = \left| \frac{i\,e^{it}}{e^{it} - \cos\theta} \right| \leqslant \frac{1}{1 - \cos\theta}.$$

Hence the contribution to I of the arc of Γ on which $0 \leqslant t \leqslant \theta - \eta$ is less in absolute value than $K\theta/(1 - \cos\theta)$ where

$$
\begin{aligned}
K &= \left| \exp nw(e^{i(\theta - \eta)}) \right| \\
&= \left| 2^n\, e^{n\theta i}\, e^{-\frac{1}{2}n(1 - i\cot\theta)\,\eta^2} \right| \{1 + O(n^{1-3\epsilon})\} \\
&= 2^n \exp\left(-\tfrac{1}{2} n^{1-2\epsilon} \right) \{1 + O(n^{1-3\epsilon})\} \\
&= o(2^n n^{\frac{1}{2} - 3\epsilon}),
\end{aligned}
$$

since $\frac{1}{3} < \epsilon < \frac{1}{2}$. And similarly for the other arc.

We have thus proved that, if $0 < \theta < \pi$,

$$I = 2^n\, e^{(n+\frac{1}{2})\,\theta i + \frac{1}{4}\pi i} \left(\frac{2\pi}{n \sin\theta} \right)^{\frac{1}{2}} \{1 + O(n^{1-3\epsilon})\}$$

and hence that

$$P_n(\cos\theta) = \left(\frac{2}{\pi n \sin\theta} \right)^{\frac{1}{2}} \sin\{(n+\tfrac{1}{2})\,\theta + \tfrac{1}{4}\pi\} \{1 + O(n^{1-3\epsilon})\}. \quad (37.4)$$

The estimate of the error term is rather poor. By a more refined argument it can be shown that $O(n^{1-3\epsilon})$ can be replaced by $O(n^{-1})$.

If $\mu > 1$, we write $\mu = \cosh\xi$, where $\xi > 0$. There are then two saddle-points $z = e^{\pm\xi}$, of which e^ξ is the higher since

$$\left| \exp w(e^{\pm\xi}) \right| = 2e^{\pm\xi}.$$

If we take C to be the circle $|z| = e^\xi$ which encloses $z = \cosh\xi$, we find that

$$P_n(\cosh\xi) = \frac{1}{2^n\pi}\, \mathscr{I}I, \quad (37.5)$$

where

$$I = \int_\Gamma e^{nw(z)}\,\phi(z)\,dz,$$

Γ being the semicircle $|z| = e^\xi$, $0 \leqslant \mathrm{ph}\,z \leqslant \pi$.

Near $z = e^\xi$,

$$w(z) = \xi + \log 2 + \tfrac{1}{2} e^{-\xi} \operatorname{cosech} \xi\, (z - e^\xi)^2 + \ldots,$$

so that the critical direction is the tangent to the semicircle Γ at e^ξ. Put $z = e^{\xi+u}$ and consider the arc $0 \leqslant t \leqslant \eta$, where $\eta = n^\epsilon$ with $\frac{1}{3} < \epsilon < \frac{1}{2}$. On this arc,

$$nw(z) = n\xi + \log 2^n - \tfrac{1}{2}n\, e^\xi \operatorname{cosech} \xi \, t^2 + O(n^{1-3\epsilon}),$$

and $$\phi(z)\frac{dz}{dt} = \frac{i\, e^\xi}{\sinh \xi}\{1 + O(n^{-\epsilon})\} = \frac{i\, e^\xi}{\sinh \xi}\{1 + o(n^{1-3\epsilon})\}.$$

Omitting a factor $1 + O(n^{1-3\epsilon})$, the contribution of the arc $0 \leqslant t \leqslant \eta$ to I is

$$I_1 = \frac{2^n\, e^{(n+1)\xi}}{\sinh \xi} i \int_0^\eta \exp\left(-\tfrac{1}{2}n\, e^\xi \operatorname{cosech} \xi \, t^2\right) dt$$

$$= \frac{2^n\, e^{(n+1)\xi}}{\sinh \xi} i \left(\frac{2}{n}\right)^{\frac{1}{2}} \int_0^\omega \exp\{-(1+\coth \xi)\, u^2\}\, du,$$

where $\omega = \sqrt{(\tfrac{1}{2}n^{1-2\epsilon})}$. By the same argument, we can replace the upper limit by $+\infty$, without altering the omitted factor. This gives

$$I_1 = \frac{2^n\, e^{(n+1)\xi}}{\sinh \xi} i \left\{\frac{\pi}{2n(1+\coth \xi)}\right\}^{\frac{1}{2}}$$

$$= 2^n\, e^{(n+\frac{1}{2})\xi} i \left(\frac{\pi}{2n \sinh \xi}\right)^{\frac{1}{2}}.$$

It is readily shown that the contribution to I of the rest of the semicircle Γ is $o(n^{\frac{1}{2}-3\epsilon})$. Hence, by (37.5),

$$P_n(\cosh \xi) = e^{(n+\frac{1}{2})\xi} \left(\frac{1}{2\pi n \sinh \xi}\right)^{\frac{1}{2}} \{1 + O(n^{1-3\epsilon})\}.$$

Again, the order term is poor, and can be replaced by $O(1/n)$.

Whilst the methods of this chapter would give complete asymptotic expansions of the Legendre polynomials, the determination of the coefficients would be difficult, and a method of Darboux [9], applicable to functions defined by a generating function, is preferable.

An account of applications of the saddle-point method to the classical orthogonal polynomials will be found in a book by Szegö [27].

CHAPTER 9

AIRY'S INTEGRAL

38. Definition of Ai(z)

The integral function

$$\mathrm{Ai}(z) = \frac{1}{3^{\frac{2}{3}}\pi} \sum_0^\infty \frac{\Gamma(\frac{1}{3}n+\frac{1}{3})}{n!} \sin\left\{\tfrac{2}{3}(n+1)\,\pi\right\}(3^{\frac{1}{3}}z)^n \qquad (38.1)$$

is known as Airy's integral, the reason for the name being that when z is real, it is equal to the integral

$$\frac{1}{\pi}\int_0^\infty \cos\left(\tfrac{1}{3}t^3 + zt\right)dt, \qquad (38.2)$$

which first arose in 1838 in Airy's researches on optics. Although it can be expressed in terms of Bessel functions of order $\frac{1}{3}$, its importance, both in applied mathematics and in the theory of the asymptotic solution of differential equations, justifies an independent discussion.

It is readily seen that $w = \mathrm{Ai}(z)$ satisfies the differential equation $d^2w/dz^2 = zw$. This equation also has solutions $\mathrm{Ai}(\omega z)$, $\mathrm{Ai}(\omega^2 z)$, where ω is the cube root $\exp(\frac{2}{3}\pi i)$ of unity. The three solutions are connected by the relation

$$\mathrm{Ai}(z) + \omega\mathrm{Ai}(\omega z) + \omega^2\mathrm{Ai}(\omega^2 z) = 0. \qquad (38.3)$$

Instead of using $\mathrm{Ai}(\omega z)$ or $\mathrm{Ai}(\omega^2 z)$ as second solution, the function

$$\mathrm{Bi}(z) = i\omega^2\mathrm{Ai}(\omega^2 z) - i\omega\,\mathrm{Ai}(\omega z) \qquad (38.4)$$

is used; it has the advantage of being real when z is real.

Let us suppose first that $z > 0$ and let us write $z = \nu^2$ where $\nu > 0$. It follows from (38.2) that

$$\mathrm{Ai}(\nu^2) = \frac{1}{2\pi i}\int_I e^{\nu^2 s - \frac{1}{3}s^3}\,ds, \qquad (38.5)$$

where I is the imaginary axis from $-\infty i$ to ∞i. A straightforward application of Cauchy's Theorem shows that the path I can be replaced by a path L (such as a pair of radii) from $\infty\omega^2$ to O and thence to $\infty\omega$. But as the integral along L converges uniformly with respect to ν on any compact set in the ν plane, the equation

$$\text{Ai}(\nu^2) = \frac{1}{2\pi i} \int_L e^{\nu^2 s - \frac{1}{3}s^3}\, ds \tag{38.6}$$

holds for all values of ν.

39. The asymptotic expansion of $\text{Ai}(\nu^2)$ by steepest descents

When $\nu > 0$, it is convenient to make the change of variable $s = \nu w$ in the integrals (38.5) and (38.6), which gives

$$\text{Ai}(\nu^2) = \frac{\nu}{2\pi i} \int_C e^{\nu^3(w - \frac{1}{3}w^3)}\, dw,$$

where C is either the imaginary axis I or a path L. There are two saddle-points $w = \pm 1$. Since L lies to the left of the imaginary axis, it seems likely that $w = -1$ will be the relevant saddle-point. If we write $w = u + iv$, we see that the level curve through $w = -1$ is a branch of the cubic curve

$$u^3 - 3uv^2 - 3u - 2 = 0,$$

and that the steepest curves through $w = -1$, given by

$$v(3u^2 - v^2 - 3) = 0,$$

consist of the real axis and a branch of a hyperbola. Figure 8 shows the level curve (full lines), the steepest curves (broken lines) and the valleys near this saddle-point. As the asymptotes of the hyperbola are $v = \pm\sqrt{3}\,u$, the hyperbolic curve is a permissible path L.

We parametrize the path of steepest descents by writing

$$w - \tfrac{1}{3}w^3 = -\tfrac{2}{3} - t^2,$$

where t varies from $-\infty$ to $+\infty$ as w describes L. We then get

$$\text{Ai}(\nu^2) = \frac{\nu}{2\pi i} e^{-\frac{2}{3}\nu^3} \int_{-\infty}^{\infty} e^{-\nu^3 t^2} \frac{dw}{dt}\, dt. \tag{39.1}$$

The rest is straightforward but rather tedious. We have to show that the conditions of Watson's lemma are applicable, and obtain an expansion of dw/dt valid near $t = 0$.

We shall not, however, complete the proof, but refer the reader to Brillouin [2]. As it stands, the argument would give the asymptotic expansion as $\nu \to +\infty$. But it is evident that the integral in (39.1) converges uniformly on any compact set in $|\nu| > 0$, $|\text{ph}\,\nu^3| < \frac{1}{2}\pi$ so that equation (39.1) is valid when

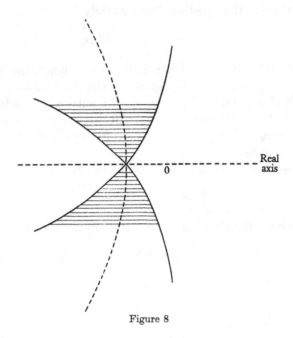

Figure 8

$|\text{ph}\,\nu| < \frac{1}{6}\pi$. The method would thus give the asymptotic expansion of $Ai(z)$ for $|z| \to \infty$ in the angle $|\text{ph}\,z| < \frac{1}{3}\pi$. We show in the next section that a simpler argument gives the required expansion in $|\text{ph}\,z| < \pi$.

We can, of course, get the dominant term immediately. It is easily shown that $dw/dt = i$ when $t = 0$. Hence when ν is positive or, more generally, when $|\text{ph}\,\nu| < \frac{1}{6}\pi$,

$$Ai(\nu^2) \sim \frac{\nu}{2\pi} e^{-\frac{2}{3}\nu^3} \int_{-\infty}^{\infty} e^{-\nu^3 t^2}\, dt$$

$$= \frac{1}{2\sqrt{(\pi\nu)}} e^{-\frac{2}{3}\nu^3},$$

and so
$$\text{Ai}(z) \sim \frac{1}{2\pi^{\frac{1}{2}}z^{\frac{1}{4}}} e^{-\frac{2}{3}z^{\frac{3}{2}}},\qquad (39.2)$$

when $|\text{ph}\,z| < \frac{1}{3}\pi$.

40. Another integral for $\text{Ai}(\nu^2)$

We return to the equation (38.5), namely

$$\text{Ai}(\nu^2) = \frac{1}{2\pi i} \int_I e^{\nu^2 s - \frac{1}{3}s^3}\, ds, \qquad (40.1)$$

where $\nu > 0$. The point in the s plane corresponding to the saddle-point $w = -1$ is $s = -\nu$. We show, by Cauchy's theorem, that the path of integration I (the imaginary axis) can be deformed into a parallel line through $s = -\nu$. All that is needed is to show that the integral along the straight line from $s = -\nu + it$ to $s = it$ tends to zero as $t \to \pm\infty$.

If we write $s = \sigma + it$, we have thus to show that

$$\int_{-\nu}^0 e^{\nu^2(\sigma+it)-\frac{1}{3}(\sigma+it)^3}\, d\sigma$$

tends to zero. The absolute value of this integral does not exceed

$$\int_{-\nu}^0 e^{\nu^2\sigma-\frac{1}{3}\sigma^3+\sigma t^2}\, d\sigma$$

$$\leqslant e^{\frac{1}{3}\nu^3} \int_{-\nu}^0 e^{\sigma t^2}\, d\sigma$$

$$< \frac{e^{\frac{1}{3}\nu^3}}{t^2},$$

which does tend to zero as $t \to \pm\infty$.

Hence, when $\nu > 0$, we may put $s = -\nu + it$ in (40.1) and integrate from $-\infty$ to $+\infty$. This gives

$$\text{Ai}(\nu^2) = \frac{1}{2\pi} e^{-\frac{2}{3}\nu^3} \int_{-\infty}^{\infty} e^{-\nu t^2 + \frac{1}{3}it^3}\, dt$$

or
$$\text{Ai}(\nu^2) = \frac{1}{\pi} e^{-\frac{2}{3}\nu^3} \int_0^{\infty} e^{-\nu t^2} \cos\left(\tfrac{1}{3}t^3\right) dt. \qquad (40.2)$$

We now observe that, when ν is complex, the integral (40.2) converges uniformly on any compact set in $|\nu| > 0$, $|\text{ph}\,\nu| < \frac{1}{2}\pi$; since $\text{Ai}(\nu^2)$ is an integral function of ν, it follows that (40.2) holds in the half-plane $\mathscr{R}\nu > 0$.

41. The asymptotic expansion of Ai(z) when $|\mathrm{ph}\, z| < \pi$

If we write $t^2 = u$ in (40.2), we obtain

$$\mathrm{Ai}(\nu^2) = \frac{1}{2\pi} e^{-\frac{2}{3}\nu^3} \int_0^\infty e^{-\nu u} \cos\left(\tfrac{1}{3} u^{\frac{3}{2}}\right) \frac{du}{\sqrt{u}}$$

when $|\mathrm{ph}\, \nu| < \tfrac{1}{2}\pi$. The conditions of Watson's lemma are evidently satisfied, and we may expand the cosine as a power series in u^3 and integrate term by term. It follows that

$$\mathrm{Ai}(\nu^2) \sim \frac{1}{2\pi} e^{-\frac{2}{3}\nu^3} \sum_0^\infty \frac{\Gamma(3n+\tfrac{1}{2})}{3^{2n}(2n)!} \frac{(-1)^n}{\nu^{3n+\frac{1}{2}}},$$

or, restoring the original variable, that

$$\mathrm{Ai}(z) \sim \frac{1}{2\pi z^{\frac{1}{4}}} e^{-\frac{2}{3}z^{\frac{3}{2}}} \sum_0^\infty \frac{\Gamma(3n+\tfrac{1}{2})}{3^{2n}(2n)!} \frac{(-1)^n}{z^{\frac{3}{2}n}},$$

as $|z| \to \infty$ in $|\mathrm{ph}\, z| < \pi$.

42. Extension of the range of values of ph z

To extend the range of values of $\mathrm{ph}\, z$, we use the identity

$$\mathrm{Ai}(z) = -\omega\, \mathrm{Ai}(\omega z) - \omega^2\, \mathrm{Ai}(\omega^2 z).$$

If we take $\omega = e^{\frac{2}{3}\pi i}$, $\omega^2 = e^{\frac{4}{3}\pi i}$, $-\tfrac{5}{3}\pi < \mathrm{ph}\, z < -\tfrac{1}{3}\pi$, we have $-\pi < \mathrm{ph}\,(\omega z) < \tfrac{1}{3}\pi$, $-\tfrac{1}{3}\pi < \mathrm{ph}\,(\omega^2 z) < \pi$, so that we can use for $\mathrm{Ai}(\omega z)$ and $\mathrm{Ai}(\omega^2 z)$ the expansion we have just found. It turns out that

$$\mathrm{Ai}(z) \sim F(z) - iG(z),$$

where

$$F(z) \sim \frac{1}{2\pi z^{\frac{1}{4}}} e^{-\frac{2}{3}z^{\frac{3}{2}}} \sum_0^\infty \frac{\Gamma(3n+\tfrac{1}{2})}{3^{2n}(2n)!} \frac{(-1)^n}{z^{\frac{3}{2}n}},$$

$$G(z) \sim \frac{1}{2\pi z^{\frac{1}{4}}} e^{\frac{2}{3}z^{\frac{3}{2}}} \sum_0^\infty \frac{\Gamma(3n+\tfrac{1}{2})}{3^{2n}(2n)!} \frac{1}{z^{\frac{3}{2}n}},$$

as $|z| \to \infty$ in $-\tfrac{5}{3}\pi < \mathrm{ph}\, z < -\tfrac{1}{3}\pi$.

If we take, as we may, $\omega = e^{-\frac{2}{3}\pi i}$, $\omega^2 = e^{-\frac{4}{3}\pi i}$, $\tfrac{1}{3}\pi < \mathrm{ph}\, z < \tfrac{5}{3}\pi$, we have $-\pi < \mathrm{ph}\,(\omega z) < \tfrac{1}{3}\pi$, $-\tfrac{1}{3}\pi < \mathrm{ph}\,(\omega^2 z) < \pi$, so that we can again use the expansion of the previous section. But the result now is that

$$\mathrm{Ai}(z) \sim F(z) + iG(z),$$

as $|z| \to \infty$ in $\tfrac{1}{3}\pi < \mathrm{ph}\, z < \tfrac{5}{3}\pi$.

We have thus obtained three asymptotic expansions for Ai(z), namely

$$F(z) \quad \text{when} \quad -\pi < \text{ph}\, z < \pi,$$

$$F(z) - iG(z) \quad \text{when} \quad -\tfrac{5}{3}\pi < \text{ph}\, z < -\tfrac{1}{3}\pi,$$

$$F(z) + iG(z) \quad \text{when} \quad \tfrac{1}{3}\pi < \text{ph}\, z < \tfrac{5}{3}\pi.$$

As the three angles overlap, it looks at first sight as though the formulae are inconsistent, but this is not so. Consider, for example, the point $z_1 = r\, e^{(\pi-\alpha)i}$ where $r > 0$, $-\tfrac{2}{3}\pi < \alpha < \tfrac{2}{3}\pi$, so that $\tfrac{1}{3}\pi < \text{ph}\, z_1 < \tfrac{5}{3}\pi$, and the point $z_2 = r\, e^{-(\pi+\alpha)i}$ so that $-\tfrac{5}{3}\pi < \text{ph}\, z_2 < -\tfrac{1}{3}\pi$. Since Ai($z$) is an integral function, it should have the same asymptotic expansions at z_1 and at z_2, since these are different descriptions of the same point in the Argand diagram. This is, in fact, the case; for, if due regard is had to phase,

$$F(z_2) = iG(z_1), \quad G(z_2) = iF(z_1),$$

so that $\qquad F(z_2) - iG(z_2) = iG(z_1) + F(z_1).$

We also had two asymptotic expansions, $F(z)$ and $F(z) + iG(z)$, both valid in $\tfrac{1}{3}\pi < \text{ph}\, z < \pi$. If we write $z = r\, e^{(\pi-\alpha)i}$ where $r > 0$, $0 < \alpha < \tfrac{2}{3}\pi$, we find that

$$|F(z)| \sim \frac{1}{2\sqrt{\pi}\, r^{\frac{1}{4}}}\, e^{\frac{2}{3} r^{\frac{3}{2}} \sin \frac{3}{2}\alpha},$$

$$|G(z)| \sim \frac{1}{2\sqrt{\pi}\, r^{\frac{1}{4}}}\, e^{-\frac{2}{3} r^{\frac{3}{2}} \sin \frac{3}{2}\alpha}.$$

Since $\sin \tfrac{3}{2}\alpha > 0$, $G(z)/F(z)$ tends to zero exponentially as $|z| \to \infty$, and so, in $\tfrac{1}{3}\pi < \text{ph}\, z < \pi$, the term $G(z)$ can be neglected. The term $G(z)$ is of importance only when $\cos(\tfrac{3}{2}\,\text{ph}\, z)$ is positive, that is when $\pi < |\text{ph}\, z| < \tfrac{5}{3}\pi$. The term $G(z)$ does not appear in the expansion when $|\text{ph}\, z| < \tfrac{1}{3}\pi$.

This change of form of the asymptotic expansion when ph z varies is another example of the Stokes phenomenon.

43. The asymptotic expansion of Ai($-z$)

If we put $z = \zeta e^{\pi i}$, we have $|\mathrm{ph}\,\zeta| < \frac{2}{3}\pi$ when $\frac{1}{3}\pi < \mathrm{ph}\,z < \frac{5}{3}\pi$. Using the formulae valid in the latter angle, we obtain

$$\mathrm{Ai}(\zeta e^{\pi i}) \sim \frac{1}{2\pi\zeta^{\frac{1}{4}}} e^{\frac{2}{3}i\zeta^{\frac{3}{2}}} \sum_0^\infty \frac{\Gamma(3n+\frac{1}{2})}{3^{2n}(2n)!} \frac{e^{-\frac{1}{4}(2n+1)\pi i}}{\zeta^{\frac{3}{2}n}}$$

$$+ \frac{1}{2\pi\zeta^{\frac{1}{4}}} e^{-\frac{2}{3}i\zeta^{\frac{3}{2}}} \sum_0^\infty \frac{\Gamma(3n+\frac{1}{2})}{3^{2n}(2n)!} \frac{e^{\frac{1}{4}(2n+1)\pi i}}{\zeta^{\frac{3}{2}n}},$$

when $|\mathrm{ph}\,\zeta| < \frac{2}{3}\pi$. And we find that $\mathrm{Ai}(\zeta e^{-\pi i})$ has the same asymptotic expansion in $|\mathrm{ph}\,\zeta| < \frac{2}{3}\pi$.

Changing the variable, we may write this result in the more useful trigonometric form

$$\mathrm{Ai}(-z) \sim \frac{\sin(\frac{2}{3}z^{\frac{3}{2}} + \frac{1}{4}\pi)}{\pi z^{\frac{1}{4}}} \sum_0^\infty \frac{\Gamma(6n+\frac{1}{2})}{3^{4n}(4n)!} \frac{(-1)^n}{z^{3n}}$$

$$- \frac{\cos(\frac{2}{3}z^{\frac{3}{2}} + \frac{1}{4}\pi)}{\pi z^{\frac{1}{4}}} \sum_0^\infty \frac{\Gamma(6n+\frac{7}{2})}{3^{4n+2}(4n+2)!} \frac{(-1)^n}{z^{3n}}$$

when $|z| \to \infty$ in $|\mathrm{ph}\,z| < \frac{2}{3}\pi$.

44. The asymptotic expansions of Bi(z)

Similar arguments can be used to deduce the asymptotic expansions of Bi(z) from the identity

$$\mathrm{Bi}(z) = i\omega^2 \mathrm{Ai}(\omega^2 z) - i\omega \mathrm{Ai}(\omega z).$$

We omit the details and merely quote the results.

When $|z| \to \infty$ in $|\mathrm{ph}\,z| < \pi$,

$$\mathrm{Bi}(z) \sim \frac{1}{\pi z^{\frac{1}{4}}} e^{\frac{2}{3}z^{\frac{3}{2}}} \sum_0^\infty \frac{\Gamma(3n+\frac{1}{2})}{3^{2n}(2n)!} \frac{1}{z^{\frac{3}{2}n}} \pm \frac{i}{2\pi z^{\frac{1}{4}}} e^{-\frac{2}{3}z^{\frac{3}{2}}} \sum_0^\infty \frac{\Gamma(3n+\frac{1}{2})}{3^{2n}(2n)!} \frac{(-1)^n}{z^{\frac{3}{2}n}},$$

where the upper or lower sign is taken according as ph z is positive or negative. This appears to introduce a discontinuity across the real positive axis. But this is not a real discontinuity, since the second term can be neglected in comparison with the first when $|\mathrm{ph}\,z| < \frac{2}{3}\pi$.

Lastly, as $|z| \to \infty$ in $|\mathrm{ph}\, z| < \frac{2}{3}\pi$,

$$\mathrm{Bi}(-z) \sim \frac{\cos\left(\frac{2}{3}z^{\frac{3}{2}} + \frac{1}{4}\pi\right)}{\pi z^{\frac{1}{4}}} \sum_{0}^{\infty} \frac{\Gamma(6n + \frac{1}{2})}{3^{4n}(4n)!} \frac{(-1)^n}{z^{3n}}$$

$$+ \frac{\sin\left(\frac{2}{3}z^{\frac{3}{2}} + \frac{1}{4}\pi\right)}{\pi z^{\frac{3}{4}}} \sum_{0}^{\infty} \frac{\Gamma(6n + \frac{7}{2})}{3^{4n+2}(4n+2)!} \frac{(-1)^n}{z^{3n}}.$$

45. An integral of Hardy and Littlewood

The integral

$$F_n(\sigma) = \int_0^{\infty} e^{\sigma x - x^n}\, dx,$$

where n is an integer greater than unity, arose in the work of Hardy and Littlewood on Waring's problem. The problem of finding the asymptotic expansion of $F_n(\sigma)$ for large values of the complex variable σ has been discussed in great detail by Bakhoom [1]. The methods used here are related to those used by Bakhoom, but simpler, since complications which arise for general values of n do not occur when $n = 3$.

CHAPTER 10

UNIFORM ASYMPTOTIC EXPANSIONS

46. The asymptotic expansions of $J_\nu(\nu a)$

We saw in sections 32 and 33 that, when ν is large and a is fixed, the function $J_\nu(\nu a)$ has, in general, an asymptotic expansion which involves an infinite series of negative powers of $\nu^{\frac{1}{2}}$, but that, when $a = 1$, the expansion involves an infinite series of negative powers of $\nu^{\frac{1}{3}}$. This change of form as $a \to 1$ was caused by the fact that two saddle-points which are distinct when $a \neq 1$, coalesce as $a \to 1$ to form a saddle-point of higher order. Evidently a phenomenon of this kind must always occur when two saddle-points coalesce in this way.

This change of form in the asymptotic expansion of $J_\nu(\nu a)$ (and, of course, of other functions which behave in a similar way) makes it desirable to have an asymptotic expansion valid uniformly in a neighbourhood of the exceptional value of the parameter.

One way of attaining this end is to go back to the differential equation satisfied by $J_\nu(\nu a)$ regarded as a function of the parameter a. By a somewhat complicated change of variables, this equation can be transformed into one which is approximately the same as the equation satisfied by Airy's integral: and one can then argue that approximately identical differential equations will have approximately identical solutions. In this way, Langer [18] obtained a uniform asymptotic approximation to $J_\nu(\nu a)$ when ν is large. The method has been extended by other authors, notably by Cherry [5] and by Olver [21], who obtained uniform asymptotic expansions for solutions of a differential equation of the form

$$\frac{d^2w}{dz^2} = \{up(z) + q(z)\}w$$

for large values of u.

The Olver type of asymptotic expansion of $J_\nu(\nu a)$ can be obtained directly from the integral definition by a method due to

Chester, Friedman and Ursell [6], which is essentially an extension of the method of steepest descents when the integrand has neighbouring saddle-points which coalesce in pairs as a certain parameter tends to some special value. We consider here the Bessel function problem, but only in the case when ν is large and positive and $0 < a \leqslant 1$, since it is possible to do so without appealing to general theorems.

47. The cubic transformation

We start with the integral

$$J_{\nu}(\nu a) = \frac{1}{2\pi i} \int_{\infty - \pi i}^{\infty + \pi i} e^{N(\sinh z - z \cosh \alpha)} dz,$$

where $N = \nu a, a = \operatorname{sech} \alpha, \alpha \geqslant 0$. This we write for convenience as

$$J_{\nu}(\nu a) = \frac{1}{2\pi i} \int_{\infty - \pi i}^{\infty + \pi i} e^{NF(z, \alpha)} dz, \qquad (47.1)$$

where $$F(z, \alpha) = \sinh z - z \cosh \alpha. \qquad (47.2)$$

The saddle-points at which $\partial F / \partial z$ vanishes are at $z = 2n\pi i \pm \alpha$, where n is any integer. In section 32 (where we discussed the case $\alpha \neq 0$) we used the quadratic transformation

$$\sinh z - z \cosh \alpha = \sinh \alpha - \alpha \cosh \alpha - t^2,$$

but, in section 33 where $\alpha = 0$, we used the cubic transformation

$$\sinh z - z = -t^3. \qquad (47.3)$$

The essential idea of Chester *et al.* [6] and Friedman [14] is to use a cubic transformation which reduces to (47.3) when $\alpha = 0$.

Let us consider then the transformation

$$F(z, \alpha) = \tfrac{1}{3} w^3 - b^2 w + c, \qquad (47.4)$$

where b and c are certain functions of α to be determined, there being no loss of generality in using a cubic polynomial without a term in w^2. If the mapping from the z plane to the w plane is to be conformal, neither dz/dw nor dw/dz can vanish in the relevant regions. Now

$$\frac{dz}{dw} = \frac{w^2 - b^2}{\cosh z - \cosh \alpha}. \qquad (47.5)$$

The numerator vanishes at $w = \pm b$, the denominator at the points $z = \pm \alpha$ in the strip $|\mathscr{I}z| \leqslant \pi$ which concerns us here. We therefore make $z = \alpha$ correspond to $w = b$, and $z = -\alpha$ to $w = -b$. This gives

$$c - \tfrac{2}{3}b^3 = \sinh \alpha - \alpha \cosh \alpha,$$

$$c + \tfrac{2}{3}b^3 = -\sinh \alpha + \alpha \cosh \alpha,$$

and so $\qquad c = 0, \quad b^3 = \tfrac{3}{2}(\alpha \cosh \alpha - \sinh \alpha).$

When $\alpha \geqslant 0$, $b^3 \geqslant 0$; we choose b to be positive when $a > 0$. If we regard α as a complex variable, b is an analytic function of α, regular in a neighbourhood of $\alpha = 0$, its singularities being at the points where $\tanh \alpha = \alpha$. The Taylor series for b near the origin is of the form

$$b = \alpha \sum_0^\infty a_n \alpha^{2n},$$

where $a_0 = 2^{-\frac{1}{3}}$, $a_1 = a_0/30$, etc.

48. Solution of the cubic transformation equation

The equation defining the cubic transformation

$$\tfrac{1}{3}w^3 - b^2 w = \sinh z - z \cosh \alpha \tag{48.1}$$

can be solved explicitly. The solution is trivial when $\alpha = 0$. But when $\alpha \neq 0$, the three solutions are

$$w_1 = 2b \sin \tfrac{1}{3}\zeta,$$
$$w_2 = -b \sin \tfrac{1}{3}\zeta + b \sqrt{3} \cos \tfrac{1}{3}\zeta, \tag{48.2}$$
$$w_3 = -b \sin \tfrac{1}{3}\zeta - b \sqrt{3} \cos \tfrac{1}{3}\zeta,$$

where ζ is the solution of

$$\tfrac{2}{3}b^3 \sin \zeta = z \cosh \alpha - \sinh z \tag{48.3}$$

which vanishes with z. This follows at once from the well-known trigonometrical solution of a cubic equation.

At $z = \alpha$, $w_1 = b$, $w_2 = b$, $w_3 = -2b$, and at $z = -\alpha$, $w_1 = -b$, $w_2 = 2b$, $w_3 = -b$. It is therefore the solution w_1 which we need. It is an odd function of z, expansible as a Taylor series

$$w_1(z) = \sum_0^\infty a_n' z^{2n+1}$$

in a neighbourhood of $z = 0$. The coefficients are complicated functions of α, the first two being

$$a_0' = \frac{\cosh \alpha - 1}{b^2}, \quad a_1' = \frac{1}{3b^2}\left\{\frac{(\cosh \alpha - 1)^3}{b^6} - \frac{1}{2}\right\}.$$

It can be shown by induction that all the coefficients a_n' are analytic functions of α^2, regarded as a complex variable, regular in a neighbourhood of $\alpha = 0$.

The function $w_1(z)$ is, in fact, of the form $zG(z^2 - \alpha^2, \alpha^2)$ where G is an analytic function of the two complex variables $z^2 - \alpha^2$ and α^2. To prove this, we observe that w_1/z is an even function which takes the value b/α when $z = \pm \alpha$, and so w_1 is of the form

$$w_1 = \frac{bz}{\alpha}\left\{1 + (z^2 - \alpha^2)v\right\}.$$

Writing $z^2 - \alpha^2 = t$ and substituting in (48.1), we obtain

$$\frac{1}{3}\frac{b^3}{\alpha^3}(1 + vt)^3(t + \alpha^2) - \frac{b^3}{\alpha}(1 + vt) = -\cosh \alpha + \frac{\sinh z}{z}. \quad (48.4)$$

But

$$\frac{\sinh z}{z} = \frac{\sinh \sqrt{(\alpha^2 + t)}}{\sqrt{(\alpha^2 + t)}} = \frac{\sinh \alpha}{\alpha} + \sum_1^\infty \frac{t^n}{n!}\left(\frac{d}{2\alpha \, d\alpha}\right)^n \frac{\sinh \alpha}{\alpha}$$

$$= \frac{\sinh \alpha}{\alpha} + \sum_1^\infty c_n t^n,$$

say. Equation (48.4) then simplifies to $H(v, t, \alpha^2) = 0$ where

$$H = v + v^2(t + \alpha^2) + \tfrac{1}{3}v^3t(t + \alpha^2) - \frac{\alpha^3}{b^3}\sum_2^\infty c_n t^{n-2}.$$

Since $c_2 = 1/5!$ and since $\alpha^3/b^3 = 2$ when $\alpha = 0$, we have $H = 0$, $\partial H/\partial v = 1$ when $t = 0$, $\alpha = 0$, $v = 2/5!$. Since H is an analytic function of v, t and α^2, the equation $H = 0$ has a unique solution v which takes the value $2/5!$ when $t = \alpha = 0$ and is an analytic function of the two variables t and α^2 in a neighbourhood of $t = \alpha = 0$.

49. Mapping by $w = w_1(z)$

The transformation $w = w_1(z)$ maps the strip $|\mathscr{I}z| \leqslant \pi$ conformally on to a certain region in the w plane, the mapping being

(1–1). The easiest way of seeing this is to consider the intermediate transformations

$$w = 2b \sin \tfrac{1}{3}\zeta, \quad \tfrac{2}{3}b^3 \sin \zeta = Z, \quad Z = z \cosh \alpha - \sinh z,$$

starting with the half-strip $\mathscr{R}z \geqslant 0$, $0 \leqslant \mathscr{I}z \leqslant \pi$. Each of these
is a univalent transformation, and therefore so also is their
resultant. The corresponding regions in the four planes are shown
in figures 9–12. The origins correspond in all four planes. The

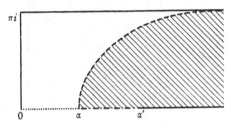

Figure 9. The z plane.

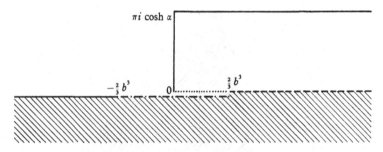

Figure 10. The Z plane.

point $z = \alpha$ is mapped on to $Z = \tfrac{2}{3}b^3$, $\zeta = \tfrac{1}{2}\pi$, $w = b$; the point
$z = \alpha'$ is mapped on to $Z = -\tfrac{2}{3}b^3$, $\zeta = \tfrac{3}{2}\pi$, $w = 2b$. The broken
curve in the z plane is the curve of steepest descent from α; its
map in the w plane is the curve of steepest descent from b. The
correspondence between areas in the four planes is indicated by
shading.

The effects of the intermediate transformations on the other
quadrants can be discussed in the same way, but the final result
is evident by symmetry. Fitting the results together, we find

that $w = w_1(z)$ maps $|\mathscr{I}z| \leqslant \pi$ conformally on to a region in the w plane and the correspondence is (1–1), though the intermediate transformations, when applied to the whole strip, are not univalent.

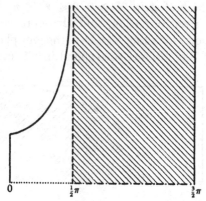

Figure 11. The ζ plane.

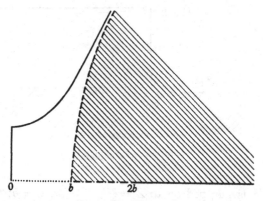

Figure 12. The w plane.

The upper boundary of the region in the w plane is the map of $\mathscr{I}z = \pi$. Now when $z = x + \pi i$,

$$w^3 - 3b^2w = -3\sinh x - 3x\cosh\alpha - 3\pi i\cosh\alpha,$$

so that the equation of the upper boundary is

$$\mathscr{I}(w^3 - 3b^2w) = -3\pi\cosh\alpha.$$

Writing $w = u + iv$, we find that the boundary is a branch of the cubic curve

$$v^3 - 3u^2v + 3b^2v = \pi \cosh \alpha.$$

It is the branch with asymptotes $v = \pm u \sqrt{3}$. The lower boundary is the reflexion of this curve in the real axis.

The curve of steepest descents through b is

$$\mathscr{I}(w^3 - 3b^2w) = 0,$$

that is $\qquad\qquad v^3 - 3u^2v + 3b^2v = 0.$

The real axis is the path of steepest ascent. The path of steepest descent is a branch of the hyperbola $3u^2 - v^2 = 3b^2$, which approaches the boundaries of the region asymptotically; it consists of the broken curve in figure 12 and its reflexion in the axis of u.

50. Derivation of the uniform asymptotic expansion

We now drop the suffix 1 and denote $w_1(z)$ simply by $w(z)$. We start with the formula (47.1)

$$J_\nu(\nu a) = \frac{1}{2\pi i} \int_{\infty - \pi i}^{\infty + \pi i} e^{N(\sinh z - z \cosh \alpha)} \, dz,$$

where $0 < a \leqslant 1$, $a = \operatorname{sech} \alpha$, $N = \nu \operatorname{sech} \alpha$, and find an asymptotic formula valid when $\nu \to +\infty$, a being kept fixed. A suitable path is the path of steepest descent through α, the dotted curve of the diagram. Changing to the variable w, we obtain

$$J_\nu(\nu a) = \frac{1}{2\pi i} \int_C e^{N(\frac{1}{3}w^3 - b^2 w)} \frac{dz}{dw} \, dw. \qquad (50.1)$$

The path C in the w plane is any curve, such as the curve of steepest descents through b, which begins at $\infty e^{-\frac{1}{3}\pi i}$ and ends at $\infty e^{\frac{1}{3}\pi i}$. We must first examine the behaviour of dz/dw regarded as a function of w.

Since $w = w_1(z)$ sets up a univalent conformal mapping of $|\mathscr{I}z| \leqslant \pi$ on to a region in the w plane, z is an analytic function of w and α^2, odd in w. Hence dz/dw is an even function of w, that is, is an analytic function of w^2. Near $w^2 = b^2$, it can therefore be expanded as a convergent power series

$$\frac{dz}{dw} = \sum_0^\infty k_n(\alpha)\,(w^2 - b^2)^n, \qquad (50.2)$$

where the coefficients are given by

$$k_n(\alpha) = \frac{1}{n!}\left\{\frac{d^n}{d(w^2)^n}\frac{dz}{dw}\right\}_{w=b}.$$

From $\qquad\qquad \frac{1}{3}w^3 - b^2 w = \sinh z - z \cosh \alpha,$

we have $\qquad\qquad (w^2 - b^2) = (\cosh z - \cosh \alpha)\frac{dz}{dw},$

and hence

$$2w = (\cosh z - \cosh \alpha)\frac{d^2 z}{dw^2} + \sinh z\left(\frac{dz}{dw}\right)^2,$$

$$2 = (\cosh z - \cosh \alpha)\frac{d^3 z}{dw^3} + 3\sinh z\frac{dz}{dw}\frac{d^2 z}{dw^2} + \cosh z\left(\frac{dz}{dw}\right)^3,$$

and so on. The second equation gives

$$\left(\frac{dz}{dw}\right)^2_{w=b} = \frac{2b}{\sinh \alpha},$$

and, from the definition of w_1, it follows that

$$\left(\frac{dz}{dw}\right)_{w=b} = +\sqrt{\left(\frac{2b}{\sinh \alpha}\right)},$$

which tends to $2^{\frac{1}{3}}$ as $\alpha \to 0$. The third equation then gives

$$\left(\frac{d^2 z}{dw^2}\right)_{w=b} = \frac{2}{3\sinh \alpha}\sqrt{\left(\frac{\sinh \alpha}{2b}\right)} - \frac{2b\cosh \alpha}{3\sinh^2 \alpha}.$$

Hence $\qquad\qquad k_0(\alpha) = \sqrt{\left(\frac{2b}{\sinh \alpha}\right)},$

$$k_1(\alpha) = \frac{1}{3b\sinh \alpha}\left\{\sqrt{\left(\frac{\sinh \alpha}{2b}\right)} - \frac{b\cosh \alpha}{\sinh \alpha}\right\}$$

and so on. The other coefficients can be determined in the same way, but there does not appear to be any simple formula for

$k_n(\alpha)$. Each coefficient is an analytic function of α^2, regular in a neighbourhood of the origin.

If we substitute the series (50.2) in (50.1), and integrate term by term, we get a series of the form

$$\sum_0^\infty k_n(\alpha) f_n(\alpha^2, N), \qquad (50.3)$$

where $\qquad f_n(\alpha^2, N) = \dfrac{1}{2\pi i} \displaystyle\int_C e^{N(\frac{1}{3}w^3 - b^2 w)} (w^2 - b^2)^n \, dw.$

In particular,

$$f_0(\alpha^2, N) = \mathrm{Ai}(N^{\frac{2}{3}} b^2)/N^{\frac{1}{3}}, \quad f_1(\alpha^2, N) = 0;$$

and the remaining functions can all be expressed in terms of Airy's integral $\mathrm{Ai}(N^{\frac{2}{3}} b^2)$ and its first derivative. Chester *et al.* [6] proved that the series (50.3) is an asymptotic expansion, uniform in α. For simplicity we shall consider here only the first term and the order of magnitude of the error.

Since dz/dw is an analytic function of w^2 and α^2, it is bounded in a neighbourhood $|w^2 - b^2| \leqslant R_1$, $|a^2| \leqslant R_2$ of the points $w^2 = b^2$, $\alpha^2 = 0$, so that $|dz/dw| \leqslant K$, say. By Cauchy's inequality, $|k_n(\alpha)| \leqslant K/R_1^n$. Hence, if $|w^2 - b^2| \leqslant \frac{1}{2} R_1$,

$$\frac{dz}{dw} = k_0(\alpha) + k_1(\alpha) (w^2 - b^2) + (w^2 - b^2)^2 \, \Phi,$$

where $\qquad |\Phi| = \left| \displaystyle\sum_0^\infty k_{n+2}(\alpha^2) (w^2 - b^2)^n \right|$

$$\leqslant \sum_0^\infty \frac{K}{R_1^{n+2}} (\tfrac{1}{2} R_1)^n = \frac{2K}{R_1^2}.$$

Thus Φ is uniformly bounded in $|w^2 - b^2| \leqslant \frac{1}{2} R_1$.

Again, if $z = x + iy$, where x is large and positive, $-\pi \leqslant y \leqslant \pi$,

$$\tfrac{1}{3} w^3 \sim \tfrac{1}{2} e^{x - iy},$$

and so $\qquad \dfrac{dz}{dw} = \dfrac{w^2 - b^2}{\cosh z - \cosh \alpha} \sim \dfrac{w^2}{\frac{1}{2} e^{x + iy}} \sim \dfrac{1}{3w}.$

Hence $\Phi = O(1/w^2)$, uniformly in α, as $|w| \to \infty$ on C. It follows that there is a constant K' such that $|\Phi| < K'$ on C for all α in some fixed interval $0 \leqslant \alpha \leqslant \alpha_1$.

We then have

$$J_\nu(\nu a) = k_0(\alpha) \frac{\text{Ai}(N^{\frac{2}{3}} b^2)}{N^{\frac{1}{3}}} + \frac{1}{2\pi i} \int_C e^{N(\frac{1}{3} w^3 - b^2 w)} (w^2 - b^2)^2 \, \Phi(w) \, dw$$

$$= \sqrt{\left(\frac{2b}{\sinh \alpha}\right)} \frac{\text{Ai}(N^{\frac{2}{3}} b^2)}{N^{\frac{1}{3}}} + \frac{1}{2\pi i} I. \tag{50.4}$$

We shall prove that, when N is large, I is small compared with the first term.

Since $\qquad\qquad w = \sqrt{(\frac{1}{3} v^2 + b^2)} + iv$

on the path of steepest descent, the square root being positive, we have

$$\left| \frac{dw}{dv} \right| = \left| \frac{v}{3\sqrt{(\frac{1}{3} v^2 + b^2)}} + i \right| = \sqrt{\left(\frac{4v^2 + 9b^2}{3v^2 + 9b^2}\right)} \leqslant \frac{2}{\sqrt{3}},$$

$$|w^2 - b^2| = \tfrac{2}{3} |v| \sqrt{(4v^2 + 9b^2)},$$

$$w^3 - 3b^2 w = -\frac{2}{3\sqrt{3}} (4v^2 + 3b^2) \sqrt{(v^2 + 3b^2)}$$

there. It follows that $\qquad |I| \leqslant \dfrac{8}{9\sqrt{3}} K'J,$

where $\qquad\qquad J = \displaystyle\int_{-\infty}^{\infty} e^{-N\psi} v^2 (4v^2 + 9b^2) \, dv$

with $\qquad\qquad \psi = \dfrac{2}{9\sqrt{3}} (4v^2 + 3b^2) \sqrt{(v^2 + 3b^2)}.$

Since the square root in ψ is positive, we may replace J by twice the corresponding integral from 0 to ∞. We consider separately the cases $b = 0$ and $b > 0$.

When $b = 0$, $\qquad\qquad J = 8 \displaystyle\int_0^{\infty} e^{-kNv^3} v^4 \, dv,$

where $k = 8/(9\sqrt{3})$. Hence $J = c_1 N^{-\frac{5}{3}}$ where c_1 is a numerica constant, and so

$$|I| \leqslant \frac{8c_1 K'}{9\sqrt{3} . N^{\frac{5}{3}}}.$$

But when $b = 0$, the first term of (50.4) is a numerical multiple of $N^{-\frac{1}{3}}$, so that I is, in this case, very small compared with the first term.

When $b > 0$, the substitution $v = bt$ gives

$$J = 2b^5 \int_0^\infty e^{-Nb^3\chi} t^2(4t^2 + 9)\, dt,$$

where
$$\chi = \frac{2}{9\sqrt{3}}(4t^2 + 3)\sqrt{(t^2 + 3)}.$$

Since χ has only one stationary value, a minimum at $t = 0$, we may approximate to J by Laplace's method when N is large and positive. This gives

$$J \sim 18b^5 e^{-\frac{2}{3}Nb^3} \int_0^\infty e^{-Nb^3 t^2} t^2\, dt = c_2 e^{-\frac{2}{3}Nb^3} \sqrt{\left(\frac{b}{N^3}\right)},$$

where c_2 is a numerical constant. But by (39.2)

$$\frac{\mathrm{Ai}(N^{\frac{2}{3}}b^2)}{N^{\frac{1}{3}}} \sim \frac{1}{2\sqrt{(N\pi b)}} e^{-\frac{2}{3}Nb^3}.$$

Hence
$$J \sim \frac{2\sqrt{\pi}}{N} bc_2 \frac{\mathrm{Ai}(N^{\frac{2}{3}}b^2)}{N^{\frac{1}{3}}}.$$

We have thus proved that, when $\alpha > 0$,

$$J_\nu(\nu \operatorname{sech} \alpha) = \sqrt{\left(\frac{2b}{\sinh \alpha}\right)} \frac{\mathrm{Ai}(N^{\frac{2}{3}}b^2)}{N^{\frac{1}{3}}} \left\{1 + O\left(\frac{1}{N}\right)\right\}, \quad (50.5)$$

where
$$N = \nu \operatorname{sech} \alpha, \quad b^3 = \tfrac{3}{2}(\alpha \cosh \alpha - \sinh \alpha).$$

The constant in the order term involves a factor b, so that when $\alpha \to 0$ the order term has to be replaced by $o(1/N)$, being, in fact, $O(N^{-\frac{1}{3}})$.

If $a \geqslant 1$, we can put $a = \sec \beta$, and apply a similar argument. When a is complex, the analysis becomes much more complicated; for details, we refer to the paper by Chester et $al.$ [6].

BIBLIOGRAPHY

[1] BAKHOOM, N. G. *Proc. Lond. Math. Soc.* (2), **35** (1933), 83–100.

[2] BRILLOUIN, L. *Ann. sci. éc. norm. sup.* (3), **33** (1916), 17–69.

[3] BURWELL, W. R. *Proc. Lond. Math. Soc.* (2), **22** (1924), 57–72.

[4] CHERRY, T. M. *Proc. Edinb. Math. Soc.* (2), **8** (1948), 51, formula 1.

[5] CHERRY, T. M. *Trans. Amer. Math. Soc.* **68** (1950), 224–57.

[6] CHESTER, C., FRIEDMAN, B. and URSELL, F. *Proc. Camb. Phil. Soc.* **53** (1957), 599–611.

[7] CORPUT, J. G. VAN DER. *Compositio math.* **1** (1934–5), 15–38; **3** (1936), 328–72.

[8] CORPUT, J. G. VAN DER. *Proc. nederl. Akad. Wet.* **51** (1948), 650–8.

[9] DARBOUX, G. *J. Math. pures appl.* (3), **4** (1878), 5–56, 377–416.

[10] DEBYE, P. *Math. Ann.* **67** (1909), 535–58; *Münch. Ber.* **40** (1910), no. 5.

[11] DINGLE, R. B. *Proc. Roy. Soc.* A, **244** (1958), 456–90; **249**, 270–95.

[12] ERDÉLYI, A. *Office of Naval Research Technical Report on Asymptotic Expansions* (Calif. Inst. Tech.) (1955). Reprinted by Dover Press, New York, 1956. See also *J. Soc. Ind. Appl. Math.* **3** (1955), 17–27; **4** (1956), 38–47.

[13] ERDÉLYI, A. *J. Soc. Ind. Appl. Math.* **3** (1955), 17–27. See also *Proc. Fourth Canadian Math. Congr.* (Toronto, 1959), 137–46.

[14] FRIEDMAN, B. *J. Soc. Ind. Appl. Math.* **7** (1959), 280–9.

[15] JONES, D. S. and KLEIN, M. *J. Math. Phys.* **37** (1958), 1–28.

[16] KELVIN, LORD (W. THOMSON). *Phil. Mag.* (5), **23** (1887), 252–5. Reprinted in his *Math. Phys. Pap.* **4**, 303–6.

[17] LAMB, H. *Hydrodynamics* (Cambridge, 1916), ch. IX.

[18] LANGER, R. E. *Trans. Amer. Math. Soc.* **33** (1931), 23–64.

[19] LAPLACE, P. S. DE. *Théorie analytique des probabilités* (Paris, 1820), 88–109.

[20] NICHOLSON, J. W. *Phil. Mag.* (16), **16** (1909), 276–7.

[21] OLVER, F. W. J. *Phil. Trans.* A, **247** (1954), 307–68. This paper contains an extensive bibliography on p. 327.

[22] POINCARÉ, H. *Acta Math.* **8** (1886), 295–344.

[23] PÓLYA, G. and SZEGÖ, G. *Aufgaben und Lehrsätzen aus der Analysis*, **1** (Berlin, 1925), 78–81, 244–9.

[24] RAYLEIGH, LORD (J. W. STRUTT). *Phil. Mag.* (6), **20** (1910), 1001–4.

[25] RIEMANN, B. *Riemann's 'Gesammelte mathematische Werke'* (2nd ed., 1892), 424–30. Reprinted by the Dover Press, New York, 1953. This paper, dated October 1863, was reconstructed from fragmentary notes by H. A. Schwarz.

[26] STIELTJES, T. J. *Ann. sci. éc. norm. sup.* (3), **3** (1886), 201–58. Reprinted in *Œuvres complètes de T. J. Stieltjes* (Groningen, 1918), **2**, 2–58.

[27] SZEGÖ, G. Orthogonal polynomials. *Amer. Math. Soc. Colloquium Publications*, **23** (1939), 215–29.

[28] TRICOMI, F. G. and ERDÉLYI, A. *Pacific J. Math.* **1** (1951), 133–42.

[29] WATSON, G. N. *Proc. Camb. Phil. Soc.* **19** (1918), 42–55. See also his *Treatise on the Theory of Bessel Functions* (Cambridge, 1922), 229–34.

[30] WATSON, G. N. *Proc. Lond. Math. Soc.* (2), **17** (1918), 116–48. In particular, p. 133. See also his *Treatise on the Theory of Bessel Functions* (Cambridge, 1922), 236.

[31] WATSON, G. N. *Proc. Lond. Math. Soc.* (2), **29** (1929), 293–308

[32] WIDDER, D. V. *The Laplace Transform* (Princeton, 1941), 277.

Reference must also be made to the following books which deal with aspects of the theory not covered in this tract.

DE BRUIJN, N. G. *Asymptotic Methods in Analysis* (Amsterdam, 1958).

JEFFREYS, H. *Asymptotic Approximations* (Oxford, 1962).

INDEX

Airy's integral, 99–106, 115–17
asymptotic expansions, Poincaré's definition of, 5–7
asymptotic power series, 7–12
asymptotic sequences, 5

Bessel functions, 34–5, 43–4, 72–82, 107–17
beta function, 60–2

confluent hypergeometric function, 17, 50–1

error function, 2, 82–7

Fourier integrals, 21–6

gamma function, 41–2, 53–8, 60–2, 70–2
    logarithm of the, 51–3

incomplete gamma function, 13–14
integration by parts, 13–26

Laplace, method of, 36–47
Legendre polynomials, 41, 94–8
logarithmic integral, 58–60

neutralizers, 24–6

parabolic cylinder function, 44–7

saddle-point method, 91–8
stationary phase, method of, 27–35
steepest descents, method of, 63–90

uniform asymptotic expansions, 107–17

Printed in the United States
By Bookmasters